STYRENE MODELING

How to build, paint, and finish realistic styrene models

COMPILED BY BOB HAYDEN

CONTENTS

evergreen
scale models

Published by Evergreen Scale Models, 18620-F 141st Ave. N.E., Woodinville, WA 98072.

Printed in China
First printing, 2000; Second Printing, 2002, Third Printing 2005

ISBN: 0-9678369-0-5

DEDICATION

To Alan Armitage, who introduced styrene
to generations of modelers

BOOK DESIGN

Luser Design, Egg Harbor, Wisconsin

COVER DESIGN

Bob Delaney, Redmond, Washington

To obtain additional copies of this book, visit your hobby shop
or call Evergreen Scale Models at (877)-376-9099.

ACKNOWLEDGEMENTS

Major thanks go to the following individuals. You'll find their names mentioned throughout the book where their work appears: Jack Burgess, Pat Hawkey of Hawx Planes, Al Kalbfleisch, James Kloek, John Nehrich, David Merriman of D&E Miniatures, Joe Morgan, Bob Robbins, Paul Scoles, Geoffrey Stippes, and Bob Steinbrunn. While what they model differs radicaly, their enthusiasm for what they build, and their willingness to share what they've learned with others make them colleagues. Thank you all!

Special appreciation is due to Brian Ellerby of Evergreen Scale Models, who saw the need for this book and provided the resources to produce it. Similar special acknowledgement is due Larry Luser, who expertly designed and produced the pages. Errors and omissions are entirely my own.

Bob Hayden
Santa Fe, New Mexico
September 2000

This immense scratchbuilt ore concentrator on Brian Ellerby's Sn3 (1/64 scale) Copper River & Yukon RR is made almost entirely from styrene sheet and strip. Brian used Grandt Line plastic window castings, and the crib work retaining walls on both sides of the structure were made from wood dowels. Photo by Paul Scoles.

Chapter 1

The Case for Styrene, 40 years later!

FOUR DECADES AGO this how-to book would have begun with a selling job. Before we could begin to show you how to build models from styrene, we would have had to convince you doing so made sense at all.

Back then, wood and metal were what "real" models were built from – hey, at least we weren't using stone! – and plastic was considered suspect. Although injection-molded plastic kits had been around for more than a decade, many established modelers saw them as toylike compared to the "true" craftsmanship of working with basswood or brass.

All that has changed, and for the better.

Styrene is now universally accepted as the modern material for building

James Kloek converted a Hasegawa 1/700 scale WWII *Essex*-class aircraft carrier kit to build a precise replica of U.S.S. *Hornet* (CV-12). He opened the hanger doors in the side of the ship, added a sheet styrene hanger deck, and scratchbuilt a hangar-deck catapult extension from styrene strip.

This diminutive 1/35 scale vehicle started as a Commanders kit of the World War II Hungarian Csaba armored car. Armor modeler Joe Morgan scratchbuilt the turret hatch, toolboxes on the rear fenders, and the antenna mounts from Evergreen styrene. The yet-to-be-painted model shows off his modifications.

models – all kinds of models. It's inexpensive, widely available, and above all, easy to work with. It brings the workability of clear soft wood to your workbench – but without the problems of wood grain, fuzz, or splinters – and combines that workability with the crisp precision of metal.

Modelbuilding with styrene is faster and easier than with any other material. Styrene is easy to fabricate using only hand tools, easy to cement, and equally easy to paint and finish to represent a variety of materials. And because molded styrene is the overwhelming choice for commercially produced model kits and detail parts, it's the perfect scratchbuilding and detailing medium to complement them.

Styrene won't warp or swell due to changes in humidity, and normal household temperature variations don't affect it. It is durable, stable, and bonds almost instantly. Styrene takes most hobby paints well, and does not require sanding or priming before applying finish coats.

Scratchbuilding with styrene

Many modelers are needlessly intimidated by the term scratchbuilding. The basic techniques for modelbuilding with styrene are extremely simple, and anyone who has built basic wood or plastic kits is ready to try a scratchbuilding project. The transition is easy and logical – from building kits, to kit modification, to building without a kit at all. The chapters that follow provide all the techniques you'll need, plus step-by-step examples.

Styrene may be combined with wood, metal, and other plastics using the appropriate adhesives. Acrylic plastic (Plexiglas, Acrylite, and other brands) can be bonded to styrene using acrylic solvent cement. This is a fast and easy way to build big models, particularly architectural models with large areas of glass. Many top professional modelbuilders use this technique for most of their work.

Even though styrene is the material of choice for most modelbuilding, it's important to understand its limitations and select other materials when they are better for your application. One example is thin, free-standing details such as hand grabs or antennas: Brass wire or photoetched parts offer superior strength for these. Another is unpainted wood structures such as pilings and timber bridges; they're easier to build and color if you use pre-stained wood.

About the Case Studies

Throughout this book we'll depart frequently from straight how-to

Bob Steinbrunn uses styrene extensively for both aircraft and ship models. His 1/48 scale Trimaster kit of the Heinkel He 162 Volksjager has been extensively modified and detailed with parts made from Evergreen sheet, strip, and rod.

Before and after! The unpainted model (above) shows the interesting mix of materials that Bob Robbins employed to build Illinois Traction System car No. 203 in O scale. The predominant material is white Evergreen styrene, augmented with brass wire and detail parts and white-metal castings (including the truck sideframes). The roof is a clear styrene part maufactured by Vane Jones. Construction of car No. 200 is identical; the model was painted by Dennis Love.

description of techniques to show you projects called "Case Studies in Styrene." The idea is not that you'll build the same model the same way the writer did – that's unlikely – but that you'll have a chance to look over the shoulders of experienced builders as they work. It's a great way for you to learn new ideas and techniques.

Now let's get going. As pioneer styrene modeler Alan Armitage wrote over forty years ago, "The things you can do with styrene are limited only by your imagination and ingenuity."

Styrene is used extensively in architectural display and planning models. Replica Scale Models of Auburn, Washington, built this 1/96 scale model (1/8" = 1') of One Pacific Tower for architects Curtis Beattie & Associates and developer Hammond Pacific Corp.

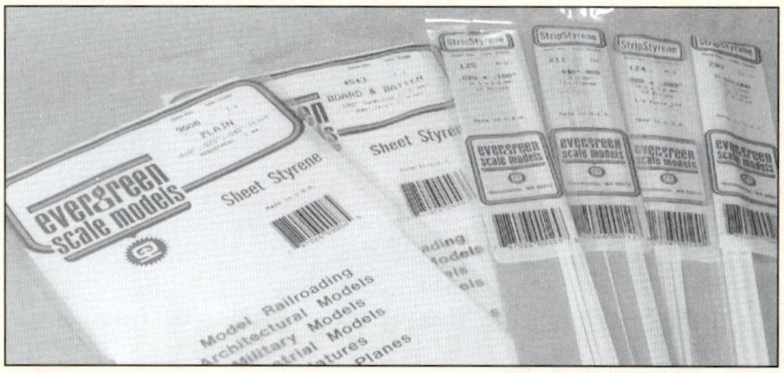

WHAT IS STYRENE, ANYHOW?

One of the most important plastics, polystyrene is found all around us – in automobiles, housewares, toys, and packaging. The basic material is derived from gases recovered in refining petroleum. It first came into widespread use in the 1950s, and one of those uses was to make model kits.

Polystyrene is a thermoplastic, which means it can be repeatedly softened with heat and formed or molded. Plastic kits and detail parts are made by heating styrene pellets (usually with a coloring agent) and forcing the taffy-like molten material into a mold, or die, under high pressure. The same process, called injection molding, is used to make everything from pill bottles to television cases and furniture.

The other common form of styrene is Evergreen strips, sheets, and shapes. This opaque white material is the same styrene used in molded plastic kits and parts. Evergreen has been manufacturing precision styrene for over 25 years.

Styrene is nontoxic, and is impervious to many chemicals, including water. It can be pliable or brittle depending on its formulation. Clear styrene is brittle and unsuitable for heat- or vacuum-forming, but makes good flat window glazing.

Ultraviolet (UV) light embrittles and degrades unpainted styrene, so it must be protected with paint for use outdoors where it will be exposed to sunlight. The much weaker levels of UV found indoors do not affect it. – *Bob Hayden*

A sharp knife, stainless-steel scale rule, and a self-healing cutting mat are all the tools you'll need for straight cuts on styrene sheet. Note how the fingers of the hand holding the rule are kept well away from the cutting edge.

Chapter 2

Equipping the Styrene Workshop

STYRENE IS EASY TO WORK WITH, and you need only a few basic hand tools. If you've built plastic kits, you probably have most of them already. All but one or two are available at well-stocked hobby shops.

In this chapter we'll go over the basic tools you'll need to work with kits and Evergreen styrene; later in the book we'll cover specialized tools when they're mentioned in connec-

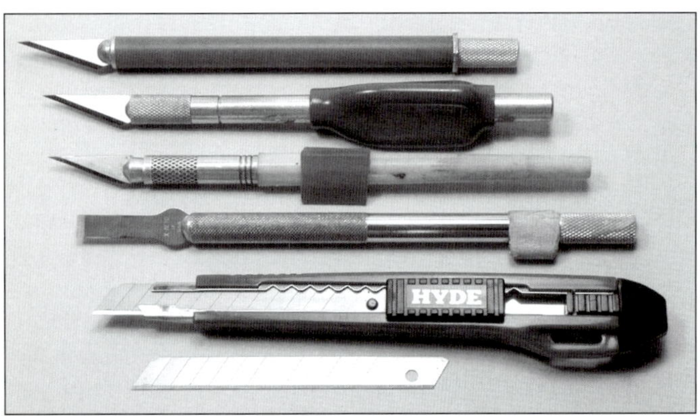

An assortment of knives collected over years of styrene modeling. Note the various additions to the three in the middle to keep them from rolling off the workbench. The bottom one is a utility knife that features segmented blades. As the tip dulls, you snap off the segment to make a new point.

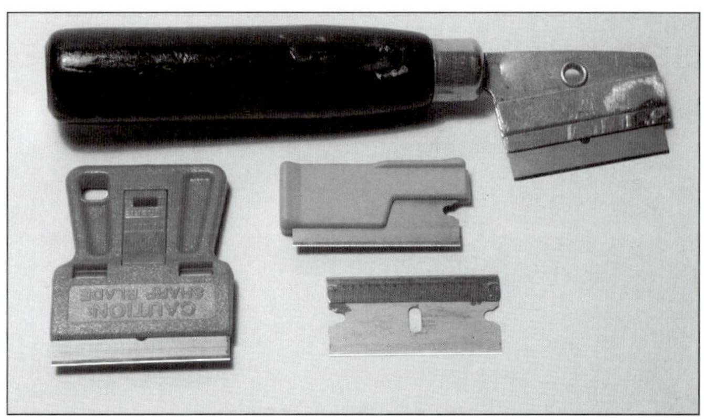

The single-edge razor blade, bottom right, is an old modeling standby. There's a variety of handles available to help you grasp them; the palette scraper on the left and the plastic gizmo in the middle came from an art-supply store, while the wood handle at top is a hardware-store item.

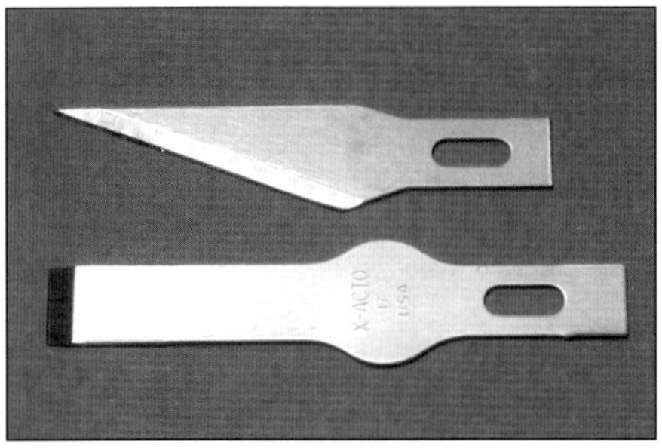

The hands-down favorite hobby knife blade is the No. 11, top. They're available in bulk packs of 100. You'll also want a few of the chisel-like No. 17 blades, bottom, for spots where the No. 11 won't fit.

tion with the specific modeling techniques that they're needed for.

Knives and blades

A sharp, thin-bladed hobby knife is the most essential styrene modeling tool. Most modelers own several, and the No. 11 blade is by far the favorite shape. It comes to a very narrow tip, and the tip is brittle; you'll break a few.

The No. 17 hobby knife blade has a 1/4"-wide chisel shape that works well for slicing away unwanted molded-in surface details on kits, such as hand grabs that you want to replace with wire parts.

For heavy cutting such as .040" or thicker styrene sheet, use a small utility knife with replaceable blades. The tips of these blades can be snapped off repeatedly to yield new, sharp points. You can find these handy knives and blades for them among the wallpaper tools at your hardware store.

Many experienced modelers swear by single-edge razor blades, which have the advantage of being inexpensive when purchased in bulk. Several types of handles are available to make them easier to hold on to and use.

Whatever type of knife you choose, keep plenty of sharp replacement blades on hand. A sharp knife is the only kind you want to pick up – it cuts cleanly and easily, requires less force, and so is safer to use.

So it's important to change blades often. I use two color-coded handles: There's a brand-new blade in the red one, and a used blade in the yellow one for jobs like opening packages, cleaning the threads of paint bottles, and other rough work.

Tweezers

Next comes a good set of tweezers with fine points. My favorite is the "MM" size, which has fine but strong tips and works with moderate pressure. It's worth owning several types for different jobs around the workbench, but you'll find that one size suits you best – buy two or three of those!

If you haven't used tweezers much, it's well worth practicing with them until you get used to their feel (like chopsticks). Tweezer points are safer than your fingers for holding small parts while gluing, as there is less chance of cement flowing between your fingers and the model parts and leaving a finger-sized surface blemish.

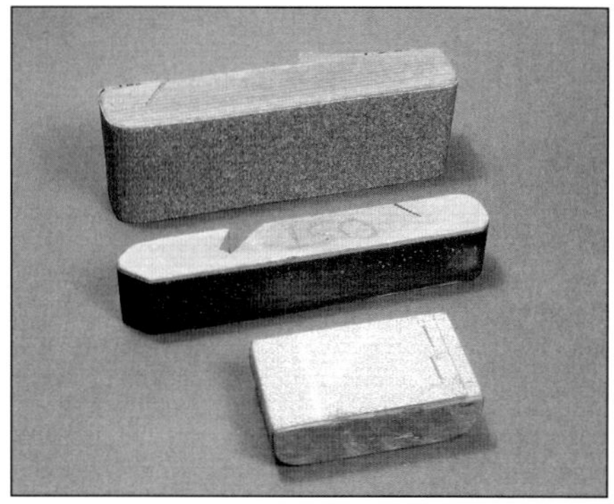

Sanding blocks are handy for truing the edges of kit parts before cementing. The middle one here is an X-acto product; the others are homemade.

Straightedge and cutting surface

You'll need a metal straightedge to guide your knife for making straight cuts in sheet stock. A 12" steel scale rule graduated for your modeling scale will do the job. An old modeling tip is to stick a strip of fine sandpaper or masking tape on the back of the rule to keep it from slipping.

The cutting surface you choose needs to be firm enough to support the styrene, but soft enough not to dull the tip of your knife blade. The self-healing cutting pads used by graphic artists are excellent for cutting styrene, and they do a good job of protecting the surface beneath them.

Another excellent cutting surface is a rectangle of Masonite or hardboard. Don't use plywood or soft wood;

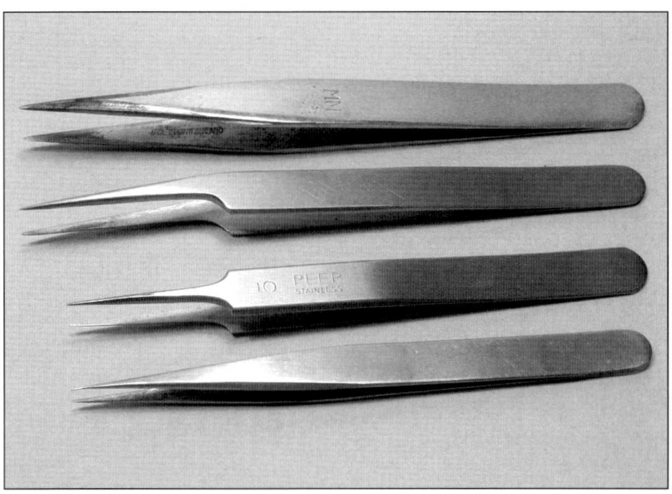

Tweezers grasp what fingers can't, and go where fingers can't reach. The well-worn MM tweezer at the top is about 5˝ long.

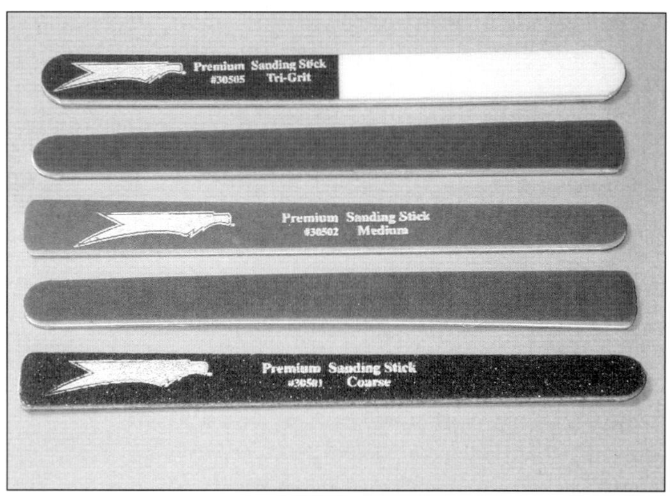

Similar to the emery boards used for manicures, these sanding sticks are surfaced with various grits of wet-or-dry sandpaper.

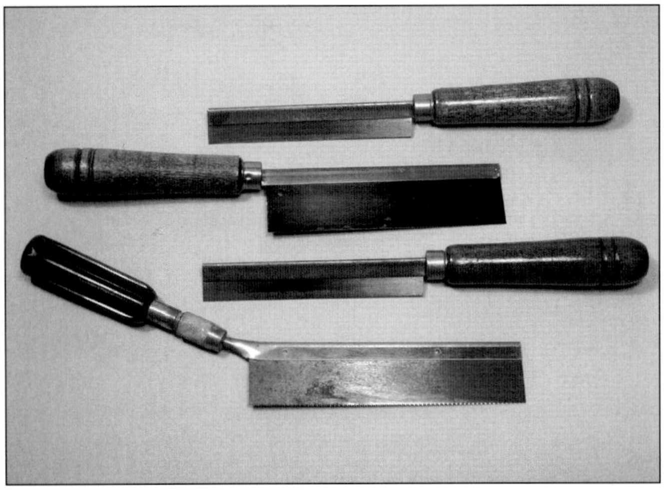

You don't need four razor saws! One fine-tooth example will suffice, but a second with a deeper blade is handy for cutting large kit parts.

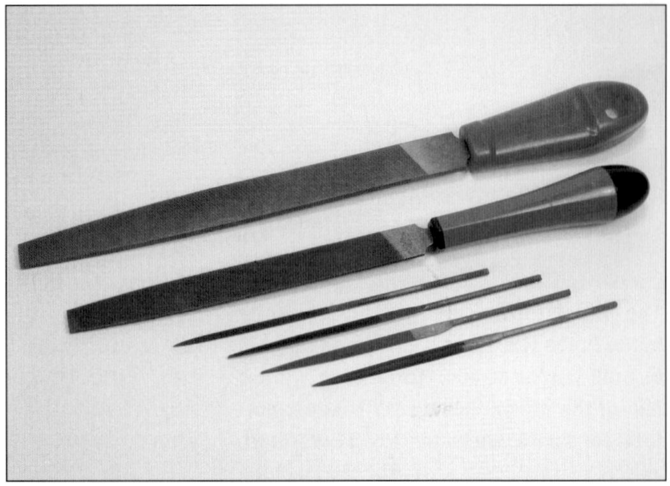

The single-cut flat mill file at the top is 12″ long, including the handle. The four needle files at the bottom have various square, round, and triangular shapes.

their grain can catch the tip of your knife blade and draw it off course.

Sandpaper, sanding blocks and sticks

Unlike wood, styrene cuts cleanly and doesn't need to be sanded to remove fuzz or grain. I keep a couple of homemade sanding blocks on the bench for finishing corners and edges, particularly on kit parts. One of these has 220-grit paper, the other 100-grit.

Several hobby companies offer sanding sticks, which are similar to emery boards but with different grades of wet-or-dry sandpaper on them. I find these handy, especially for working with kit parts (they're great for smoothing out sprue-connection marks), and I also keep a handful of standard emery boards around the workbench for rough-shaping jobs.

Saws and files

Styrene cuts easily with most fine-tooth hand saws. You should have a fine-tooth hobby razor saw for cutting thick injection-molded kit parts and styrene sheet and strip thicker than .080″.

Better-quality razor saws have cutting blades that can be replaced. Try reversing the blade on one of these so the teeth cut on the pull stroke. You'll find that cuts are easier to start and easier to control.

Occasionally a jeweler's saw comes in handy for cutting thick kit parts. You have to cut slowly, otherwise the styrene can melt. Really heavy sections call for a hacksaw with a fine-tooth blade.

Power saws are useless with styrene, in fact, they can be downright dangerous because the heat from the blade melts the styrene and makes it stick and jam.

With styrene, a sharp file actually yields a better surface finish than sanding, especially when squaring joint edges. You'll need a relatively coarse 6″ to 12″ single-cut mill file for truing edges and shaping parts. Once you get used to it, you'll find it works a lot faster and better than the tiny needle files sold for hobby use.

Needle files are good for shaping openings and making small parts. Most needle file sets include one or

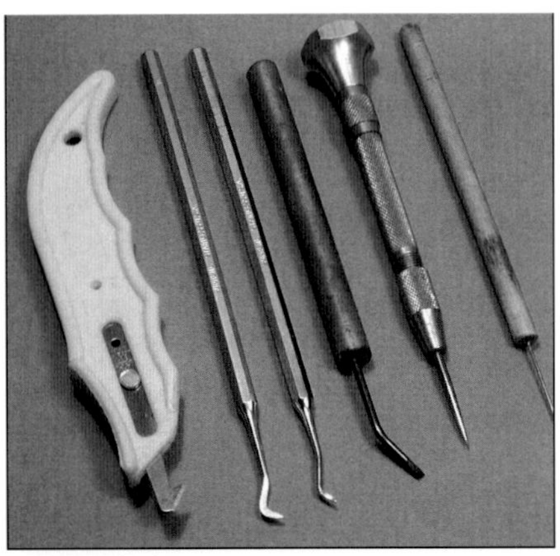

Scribers, left to right: a special tool for scoring acrylic sheet, two dental picks, a homemade groove scriber, a compass point in a pin vise, and a darning needle mounted in a simple 1/4" dowel handle.

A vernier caliper (this one is graduated for HO scale, plus decimal inches) and a couple of metal machinist's squares will bring accuracy to your modeling.

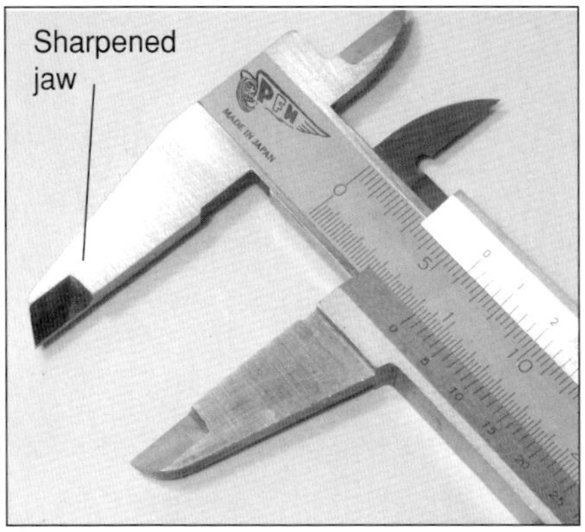

The tip of the fixed jaw of this caliper has been ground to a sharp point. This makes it ideal for transferring measurements to styrene sheet by light scoring.

two shapes with smooth edges that work well for filing inside corners such as door and window openings. Keep in mind that files are cutting tools, and they get dull and wear out. If you do a fair amount of modeling, be sure to replace your files from time to time.

Squares, scribers, and calipers

You'll need one or more small machinist's squares for laying out rectangular parts, cutting parts to length, and cementing square corners. I like the 4" size, which is heavy enough to hold itself in place while assembling right-angle joints.

A scriber, which can be a needle or a sharp compass point chucked in a pin vise, is useful for marking or scoring styrene sheet. A chisel-point scriber made from a dental pick can be used to make grooves, cut panel lines, or engrave heavy wood grain.

An inexpensive dial or vernier caliper is an investment that can greatly improve the accuracy of your modelwork. You'll use it to check the sizes of strips and sheets, and by sharpening one jaw you can use it as a marking gauge for transferring measurements to your work.

Sprue cutters

Injection-molded parts come attached to plastic trees (also called sprues, or mold gates), and must be removed before assembly. The quickest way to do this is with diagonal cutting pliers. Once a part is removed from the tree, trim away the sprue stub with a sharp No. 11 blade.

Several companies offer special pliers for the job; it's worthwhile to reserve one pair exclusively for sprue cutting, since using them for cutting wire will dull them too much to cut styrene cleanly.

Other tools

Add a set of numbered twist drills appropriate to the size of models you'll be building, a pin vise to chuck them in, and a small bench vise, and you've got all the implements you'll need to build styrene kits, modify kits, and scratchbuild models to your heart's content. The only other category of tools you'll need are those for painting and finishing, and we'll describe them in the chapters on those topics.

Of course, you can own as many tools as you have room to store: I do! But as an old torpedoman's mate once told me on a pistol firing range, "It ain't the gun, it's the gunner!"

Styrene cements and bonding techniques

Unlike wood or other porous materials which must be glued, styrene parts are joined by bonding with a solvent, which dissolves and fuses the mating surfaces. Suitable solvent cements include Testors Liquid Plastic Cement, Tenax 7R, and Ambroid Pro-Weld.

Testors, which comes with a brush mounted in the bottle cap, is the most readily available and easiest to use, though it is not the fastest drying. Tenax 7R dries slightly faster, but does not seem to form as strong a joint as Testors. I suggest you keep at least two "speeds" of solvent on your bench.

An even faster drying solvent is methyl-ethyl-ketone (MEK), which is available at paint and hardware stores in pint containers. MEK is a powerful, highly volatile solvent, and adequate ventilation is a must to ensure you don't inhale its dangerous

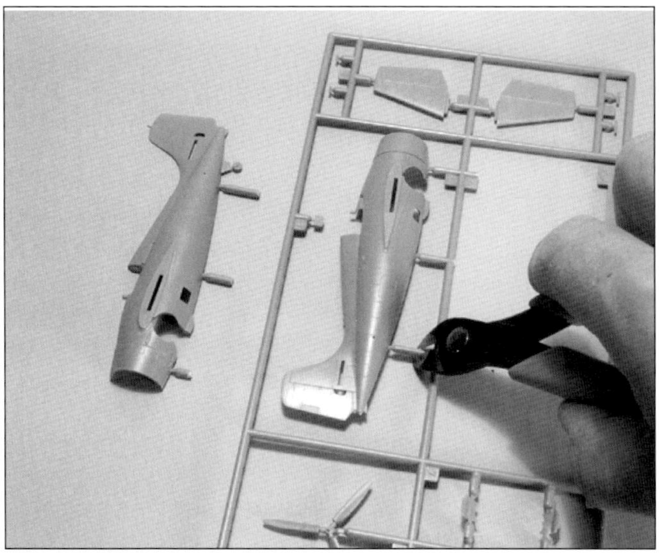

Every plastic kit comes with instructions to cut, never twist or break, the parts from the trees. Here's how. Start by cutting the tree near the part, then trim the stub away with a knife.

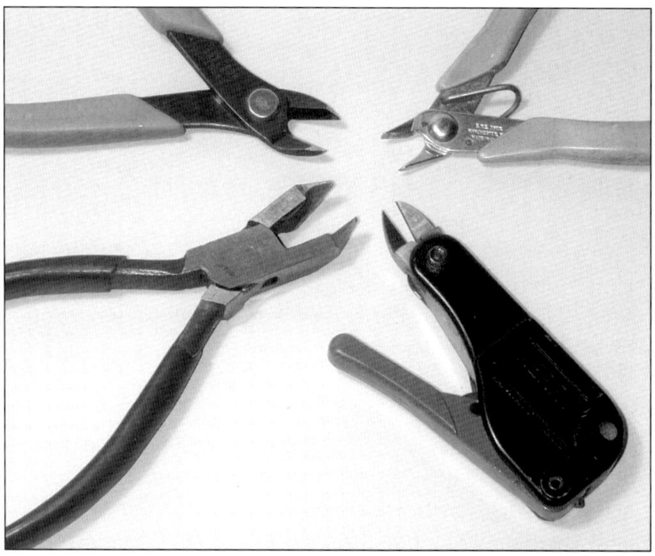

The three tools at top and left are standard wire-cutting pliers; the odd-looking device at bottom right is a tool offered specifically for cutting sprues.

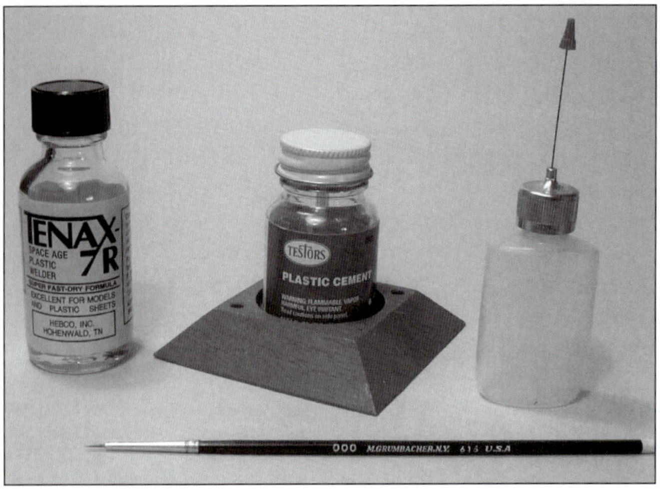

Common solvents for styrene include Hebco's Tenax 7R and Testors Liquid Plastic Cement. The most popular applicators are the fine paintbrush and the needlepoint applicator bottle. The base on the Testors bottle helps prevent knocking it over.

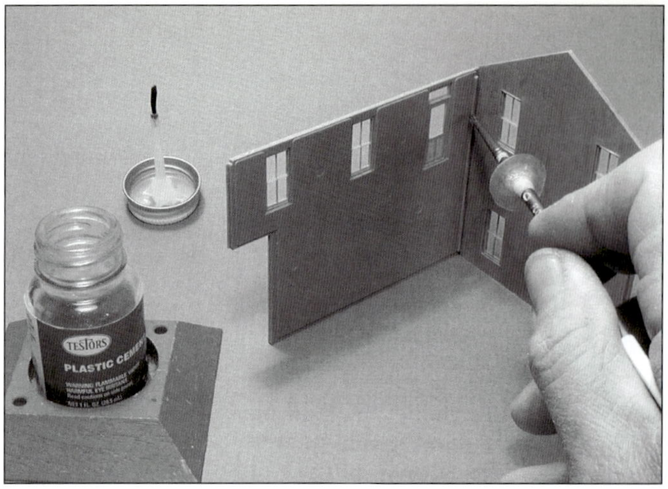

Wherever possible, solvent should be applied to the inside or back of the model parts. Note the use of a paintbrush instead of the relatively coarse applicator brush provided in the Testors bottle.

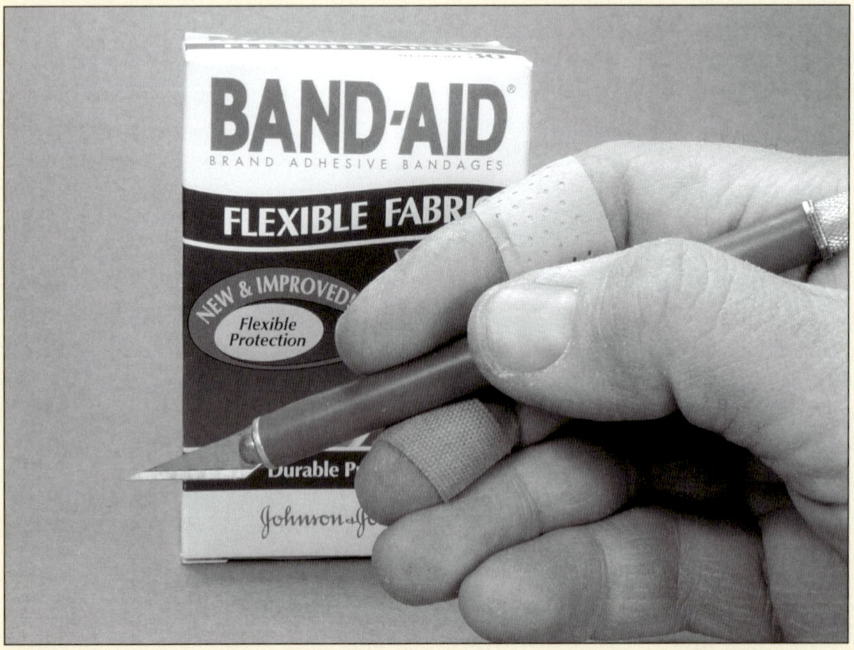

SAFETY AT YOUR WORKBENCH

There isn't much outright danger associated with styrene modelbuilding, and styrene itself is an extremely safe material. Some of the tools and materials we use can be hazardous if handled carelessly. Here are a few things to keep in mind.

Cutting tools

Hobby knives are as sharp as surgical scalpels (in fact, some modelers use scalpels), and they can cut you as well as the styrene. Teach yourself the habit of cutting with the blade pointed away from your fingers whenever possible, so a slip doesn't result in a wound.

Modeling knives with round handles can roll off your workbench, and they always seem to fall to the floor point down. It's a natural reaction to reach out with your foot to break the fall, but you'll want to practice letting objects drop from your bench without doing so.

Sharp tools are safe tools. Using a dull knife or saw means you have to apply extra pressure to force it to cut, and that extra pressure can lead to a slip and a nasty cut.

Power tools don't work well with styrene, because their high speed melts the material and causes the cutting blade to gum up and jam.

Finally, keep a box of Band-Aids on your workbench. I do.

Solvents and paints

All solvents used for cementing styrene are volatile chemicals that deserve respect. Start by reading and following the directions on the container.

Keep solvent cements safely away from fire, open flames, or other ignition sources, and make sure you have adequate ventilation through your workbench area. (If you can detect a strong solvent or paint smell, you don't have adequate ventilation!) Cap the solvent bottle whenever possible, including between applications.

Solvents are harmful if they contact your eyes, and should not be swallowed. That goes for paints and super glue, too.

Super glue requires special care, since it was originally formulated to bond skin in surgical procedures. Most experienced modelers have glued their fingers together or to the model with the stuff at least once. Acetone or the special super-glue debonders sold in hobby shops will dissolve the glue – but very slowly. – *Bob Hayden*

fumes.

Styrene-to-styrene joints are made by positioning the parts and applying a small amount of solvent to the mating area. Use a small brush (No. 00 or 000) and apply the cement sparingly – only a little is required to make strong joints. The solvent will be drawn into the joint by capillary action, softening the mating surfaces so that a bond as strong as the styrene is formed.

Solvent cements attack the surface of styrene and cause wrinkles and blemishes called "crazing." Evergreen styrene sheet and strip is more resistant to solvent than the plastic in injection-molded kits. MEK will severely craze molded parts, though it works well on sheets and strips.

The key to cementing styrene parts is to use small amounts of solvent. Too much, or excess solvent spilled on the parts, will ruin them. Thin styrene sheet parts are especially sensitive, and too much solvent can soften them all the way through, resulting in warps or sink holes.

It's altogether too easy to tip over the solvent bottle and ruin the model you're working on. (I've spilled more solvent than I've ever used to assemble models!) You can reduce the danger by mounting the bottle in a stand made from a block of wood or cementing it to a 3" square of cardboard or heavy styrene sheet with super glue.

The thick styrene cement in tubes is not recommended as it is difficult to use, slow drying, and will warp thin styrene sheets.

Solvent applicators

You'll want to earmark one paintbrush for exclusive use as a cement applicator, since repeated dips into the solvent quickly dry out the hairs and make the brush useless for painting. Wrap some tape around the handle to remind you that this brush is not for paint.

There are several alternatives to the brush. Fine needlepoint applicator bottles work well for assembling kit parts, especially big seams such as ship hulls and aircraft fuselages. Wherever possible, use them to apply solvent to the inside or back of the seam, where excess won't be visible.

On large parts, the pros often use a hypodermic syringe to apply solvent along a joint. This technique is fast, but practice is required to avoid using too much cement. When laminating large parts, use an eyedropper to apply the solvent quickly to both surfaces, then press together immediately.

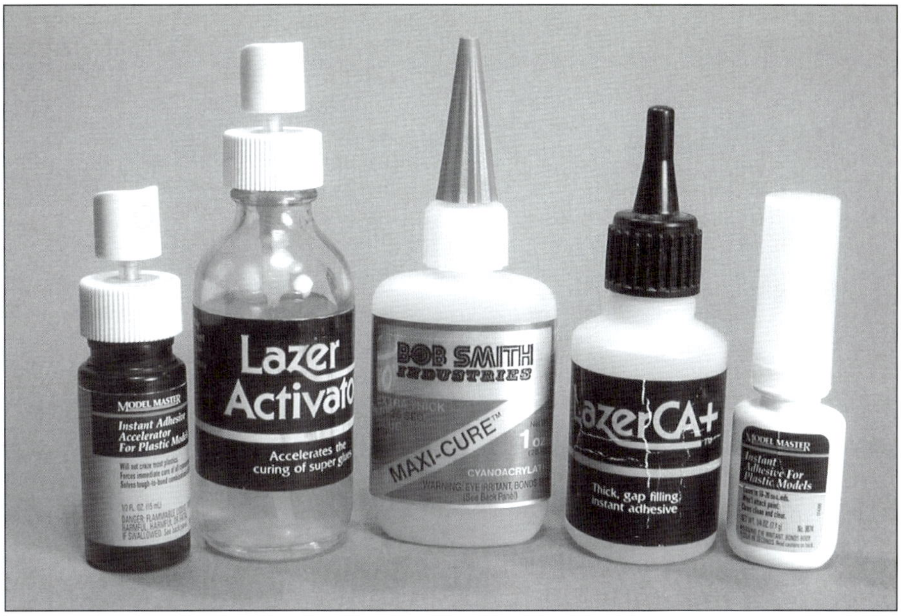

Super glues come in various formulations that affect viscosity and setting speed. It's worth having a thin, fast-setting type and a gap-filling variety on your workbench. The spray bottles of accelerator on the left force the glue to set instantly.

Bonding styrene to other materials

Solvent cements won't bond styrene to other materials or to painted styrene surfaces. For joints between styrene and wood or metal, use super glue (also called CA, ACC, or alpha cyanoacrylate). Where strength is important, use fast-setting two-part epoxy.

Gap-filling super glue is best for wood-to-styrene joints. Wire parts, such as handrails and grab irons, should be pressed into snug holes and glued with super glue, preferably from inside the model wherever possible.

Water-soluble contact cement is safe, easy, and fast for flat styrene-to-wood laminating jobs. Artist's acrylic matte medium will also work on large surface areas if the back side of the styrene is roughened up with sandpaper.

MODELING AROUND CHILDREN

Small children can't recognize or understand safety hazards, so it's your responsibility to keep modeling materials out of their reach. Store your tools, paints, and solvents in a locked desk or closet anytime you can't be present to monitor them.

Kids love to rummage through wastebaskets, and a discarded hobby knife blade can result in a nasty gash. If you have small children around, wrap the blades with several layers of masking tape or make a special effort to throw them out where the kids won't find them. – *Bob Hayden*

Painted styrene cannot be glued with solvent cement, as the solvent will not adhere to paint. If the edge of a prepainted part can be masked before painting or the paint scraped off, solvent may be used. Otherwise, painted styrene must be glued with the same adhesives used for styrene-to-metal bonds.

Tom Piccirillo built the body of this O gauge steeple cab electric locomotive from Evergreen scribed plastic siding, but used a variety of other materials, including wood, plywood, and metal castings, to complete the project.

Chapter 3

Working With Styrene Sheet and Strip

THE PHENOMENAL EASE of cutting sheet styrene is a big part of what first attracted modelers to the material. It has no grain, so it works the same in all directions and will not split. Because of this uniformity, you can rapidly fabricate parts using a simple, easy-to-learn technique called "score and snap."

Cutting styrene sheet

The thinnest styrene sheet – .005" and .010" – can be cut with shears or scissors, but it tends to curl. You're better off using a sharp hobby or util

The essential skill for working with styrene sheet and strip is the score-and-snap technique. To score, lightly draw a sharp hobby knife blade along the desired cut line. Note how the machinist's square is being used to ensure a square cut.

Gently bend the sheet away from the score and it will snap cleanly along the scored line. It's rarely necessary to cut more than a fourth of the way through the sheet. You may be pleasantly surprised at the precision of the snapped edge.

TAKING STYRENE SERIOUSLY: BOB ROBBINS' POWER STATION BOILERS

Traction modeler Bob Robbins of Fanwood, New Jersey, says Evergreen styrene is his all-time favorite modeling material. For him, the hobby is all about scratchbuilding with sheet and strip styrene, and his O scale (1/48) stationary boiler unit for a power plant is ample proof.

Here's the rectangular brick boiler foundation and firebox, partially assembled. The basic reinforced styrene-sheet box is sheathed with thousands of individually cut styrene "bricks," each spaced .010″ apart with recessed strip-styrene mortar.

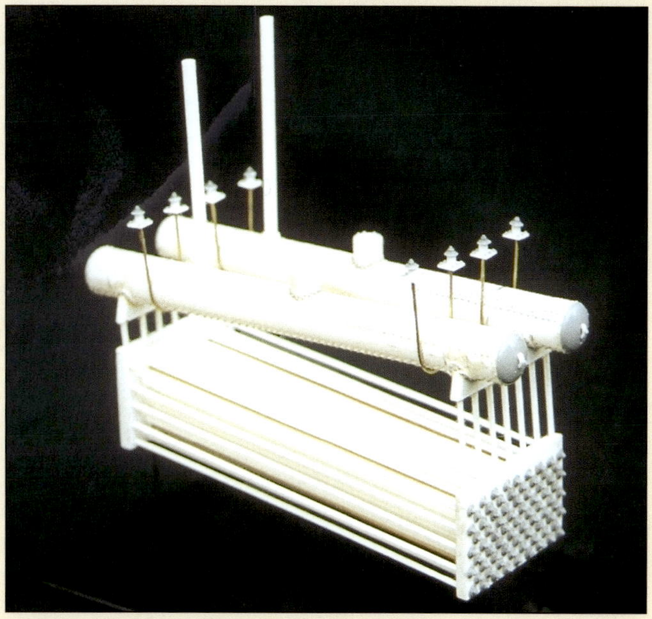

Even though the watertube boilers would eventually be mostly hidden by brick, Bob modeled them fully. Note the four .060″ brass rods installed between the styrene boiler tubes to add strength, and U-shaped brass rods that will support the assemblies from the overhead structural frame.

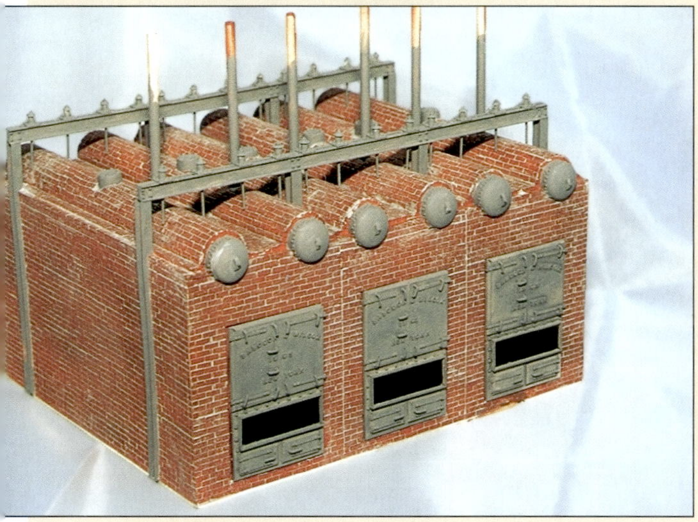

Almost completed, this is just one of eight Babcock & Wilcox boilers for a model of an electric power-generating station. The unit is built completely of styrene and has over 13,000 parts. The doors feature raised lettering: "Babcock & Wilcox, 1905, New York." The black openings are where the chain stokers will be installed.

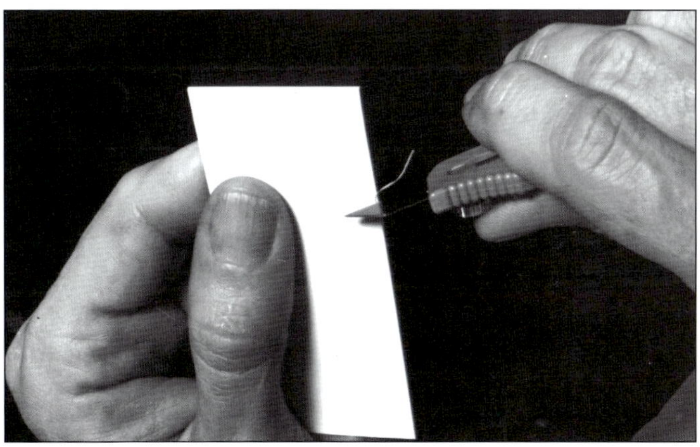

If scoring leaves a raised burr along the snapped edge, hold the hobby knife blade perpendicular to the surface to scrape it away. The extra plastic should curl cleanly away from the edge as shown here.

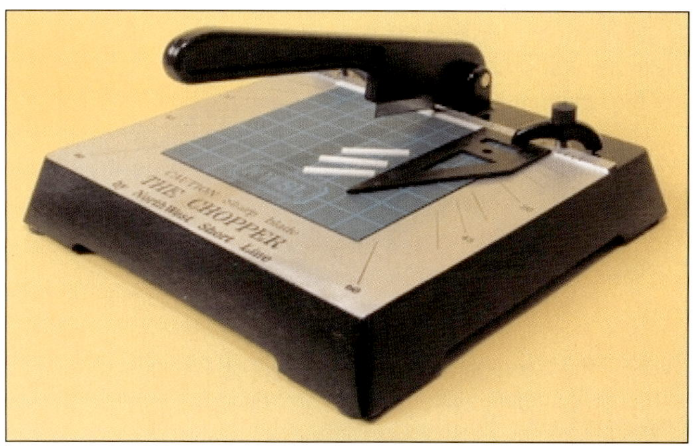

NorthWest Short Line's Chopper cutting tool is ideal when you need to make a number of identical strip styrene parts. The tool mounts a replaceable single-edge razor blade, and comes with a replaceable cutting mat.

Here's a typical strip-styrene project. Geoffrey Stippes scratchbuilt this Sn3 (1/64 scale) stock car using a variety of sheet and strip sizes. The black detail parts are commercial castings.

ity knife. On sheets up to .015" you can cut clear through with one or two passes.

Sheets .020" thick and on up call for the score-and-snap method. Align your metal straightedge with the cut

STYRENE ASSEMBLY TIPS

by Jack Burgess

TIP NO. 1
CUTTING PARTS OVERSIZE

Whenever possible, it is better to cut small pieces of styrene too long, then trim them to size after they have been bonded to the model. For example, rather than cut the letterboard for a boxcar exactly and bond it to model, cut it over-length, bond it in place, and then cut it flush with the ends.

This technique works well for making rafter tails which, regardless of how carefully you work, seem to vary in length from one end of a structure to the other. Cut a number of extra-long pieces with the correct angle on one end, glue in place, and then cut the excess flush with the edge of the roof. Flush-cutting tweezers or sprue cutters are excellent for this purpose.

This technique can sometimes be used for pieces that don't overhang an edge. For example, when adding trim around a window, cut the side trim pieces over-length, then bond the first one in place with just a dot of cement near the center of the piece. Apply just enough solvent to keep the piece from moving, but not enough that capillary action bonds the ends.

Add the trim piece for the opposite side of the window in the same manner. Once they have dried for a couple of minutes, cut off the excess at the top and bottom with a single-edge razor blade, then bond them in place permanently.

Now cut the top trim piece over-length and bond it in place with a dot of solvent. Cut the ends flush with the outside edges of the side trim pieces and finish up with the bottom trim piece.

This approach can be used in many situations. When making rectangular concrete footings for a bridge or water tank, I cut two opposite sides exact size and the other two sides oversize. Glue the box together with the tops of all four sides flush, then add an oversize top piece.

To strengthen the assembly for the handling needed in the final step, reinforce the inside corners with bracing. When the solvent has dried, file all of the outside edges smooth, rounding all the corners slightly in the process.

TIP NO. 2
ALIGNING PARTS BY FEEL

Neat construction requires square joints and careful workmanship. There are a number of techniques to help achieve

Jack's HO (1/87 scale) model of the Yosemite Valley Railroad's twin tanks at Bagby is 99 percent styrene. The tanks were formed from PVC pipe covered with scribed styrene which had been pre-curved by dipping it in hot water to better conform to the diameter of the pipe.

these results. For example, when bonding a long narrow strip to a flat sheet it is easy to lightly bump or move the strip, either when you are ready to bond it in place or before it dries.

To avoid this problem, lightly scribe a line to locate one side or edge of the strip. You can then slide the strip up to the line – the microscopic ridge created by scribing will let you feel when the strip is in position. Hold it in place with the tip of an X-acto knife and bond it in place with solvent.

TIP NO. 3
ALIGNING PARTS WITH GUSSETS

If the interior of a model will not be seen, it is typical to add gussets to reinforce the inside joints. Rather than add these gussets after the joints have dried, install them before assem-

line, and using your sharpest blade, make a shallow score along the guide.

Your first pass with the blade should be very light, with your emphasis on making a straight line that only lightly breaks the surface. This will provide a guide for your second, and if necessary, third passes.

Go back over the line with additional light passes to score the styrene for breaking. A couple of light passes

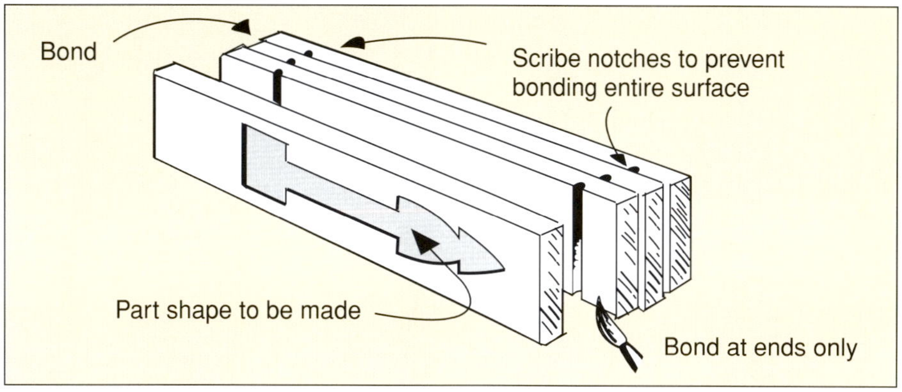

The rafter tails on Angelo's (Chapter 4) were cut over-length, glued in place, then cut even with the roof using flush-cutting tweezers. The building is resting upside down to help ensure the rafter tails are cut perpendicular to the floor.

bling the joints to help ensure that everything goes together square.

I cut the gussets from .040″ styrene in the shape of a right triangle and as large as possible. Be sure that the right angle is precise. Glue the gusset to one of the walls, flush with the edge. Once the gusset dries, bond the mating wall in place tight up against the gusset. This is an easy way to ensure that the resulting joint is glued together true and square.

TIP NO. 4

ALIGNING PARTS HORIZONTALLY

When adding identical details along a wall or panel, remember that a viewer can more easily discern when items are not level with each other than when they are not equally spaced. Since the holes cut for window castings can be a little larger than needed since the window trim will cover the excess, it is easy to end up with them crooked.

To eliminate this problem, clamp a straightedge parallel to the bottom of the side at the bottom of the window trim before gluing any of the windows in place. This will ensure that they are all even and square with the bottom of the building if they are all touching the straightedge. To keep from accidentally bonding the straightedge to the building, apply a drop of cement at the top of each window first to hold them in place, then remove the straightedge before bonding the other edges.

To make sure that the tank railings would end up square, Jack made jigs for assembling them around the tanks.

TIP NO. 5

BRACING FOR SCRIBED-STYRENE WALLS

Since styrene is unaffected by moisture, it doesn't need the interior reinforcing typically required for wood models. I rarely reinforce plain sheet styrene models. However, large expanses of milled or scribed styrene siding can have a tendency to warp or bow over time.

Adding square reinforcing styrene strips (.080″ or larger) can help, but I prefer to use .040″ styrene on edge. For example, a 3/8″- to 1/2″-wide strip of .040″ styrene glued on edge (perpendicular to the scribing) will easily reinforce a piece of scribed styrene several inches wide. If the piece is more than a couple of inches high, add another reinforcing strip parallel with the first and an inch or so away.

The tank footings were built up from sheet styrene. Grandt Line nut-bolt-washer castings were added where appropriate. Rather than drilling holes, Jack cut the castings off flush at the bottom of the washer and simply glued each in place.

should be adequate. You do not need to cut more than a fourth of the way through.

Grasp the sheet on both sides of the score and bend until it snaps. If it resists, deepen the score. On heavy sheet, .060" and over, you may have to clamp the sheet in a vise or bend it over the sharp edge of a table or workbench.

The sheet will break cleanly with only a slight burr on the top edge. Use the edge of the knife blade as a scraper to remove the burr. You may occasionally have to true the snapped edge with a file if it will be visible or part of a joint.

Practice this technique before you start cutting parts you want to keep. Scoring and snapping a 1"-wide strip to make a handful of triangular right-angle gussets is good practice, and they'll be sure to come in handy for projects later on.

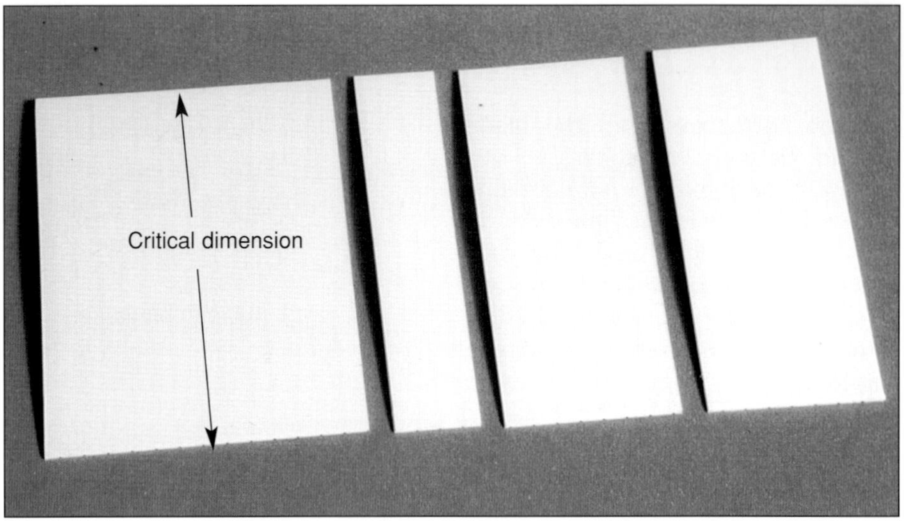

A simple technique for ensuring parts are precisely the same length or height is to cut a piece of siding or sheet stock as *wide* as the dimension that must be precise. Then slice off pieces as needed to fabricate the parts you need.

Cutting window and door openings

Making window and door openings is fast and easy. If the opening is not near the edge of the sheet, the basic score-and-snap method works on sheets up to .040" thick. Simply score deeply using several passes, then carefully flex the opening on all four sides.

Another approach is to score the outline of the window or door, then score an "X" connecting the corners. Remove the material from the openings by flexing and snapping out the four triangular scrap pieces. This places less stress on the part.

Since styrene has no grain, there is less risk of splitting the siding when cutting near an edge than there is with wood. A method for openings near the edge of the part is to drill a pilot hole and saw out the opening with a jeweler's saw. A pillar file with two smooth edges is excellent for filing square corners.

See Chapter 4 for Jack Burgess' technique for making openings with a nibbling cutter.

Cutting strip stock

Styrene strips are cut to length using the same method as for sheets. On thicker strips, score the strip and snap it. This is faster, cleaner, and actually more accurate than sawing.

Thin strips are easy to cut with a guillotine motion and a single-edge razor blade. A strip cutter such as the NorthWest Short Line "Chopper" is great for such cutting, and for making duplicate parts.

Making identical parts

To cut and file a number of identical parts to shape, stack the number of sheet or strip pieces needed and bond them together at the edges or ends. Before bonding, scribe or file grooves in the stock outside the area where the parts will be made so solvent won't bond the parts themselves. Then shape the stack of parts and cut off the bonded ends or edges as the final step.

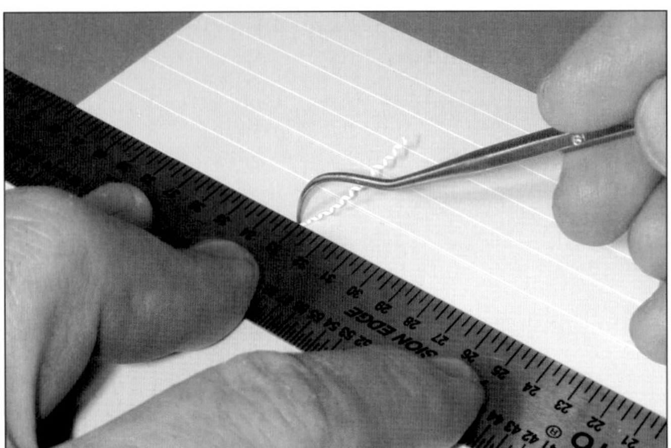

This Micro-Mark styrene scriber is made from a dental tool. Note how it removes a clean spiral chip of material as it cuts a V-shaped groove. Uniform pressure yields identical depth.

Applying solvent to Evergreen styrene sheet and strip will turn the exposed surface slightly glossy. These "solvent stains" will be hidden completely by your first coat of paint.

BUILDING A PERSONAL STYRENE STRIP INVENTORY

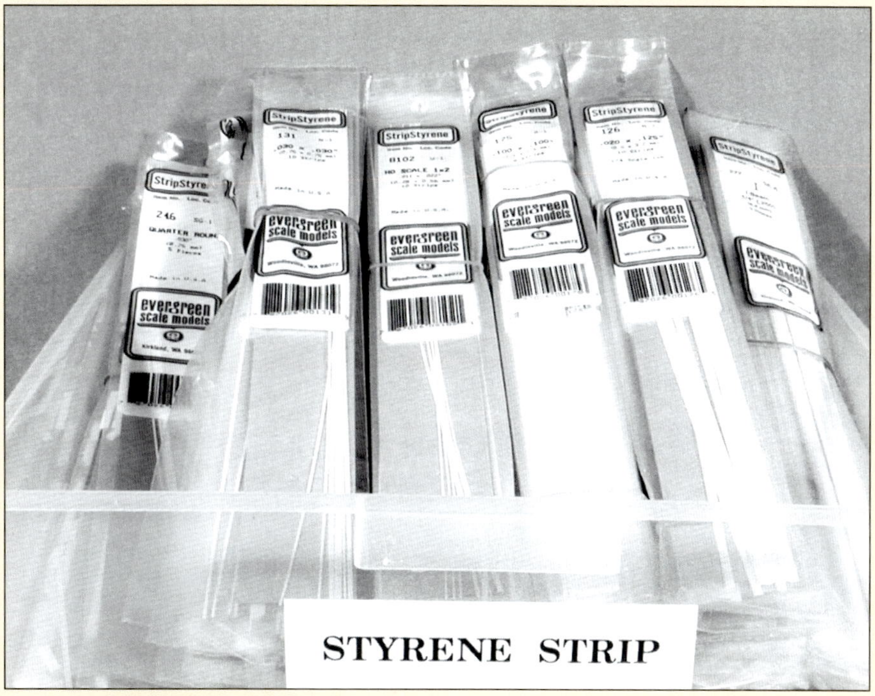

When Evergreen styrene strips first appeared in hobby shops almost 25 years ago, it was like a dream come true. Before Evergreen, the first step in a styrene modeling project was to cut your own strips from sheet stock. It was tedious, and the strips weren't always precise. Most modelers preferred stripwood, mainly because you could depend on accurate dimensions.

It didn't take long to figure out that the Evergreen strips were far more accurate than wood! I laid in a supply of every size available at the time, and I've been expanding my inventory every since.

The tip is this: Instead of buying packages of Evergreen strips only as you need a particular size, start out by picking up a couple of dozen sizes. Then expand your assortment as you tackle various projects.

Before long you'll have most — if not all — of the sizes you need to start a project. And, if you remember to replenish your inventory, you won't find yourself running out of styrene on a Sunday evening when your hobby shop is closed! – *Bob Hayden*

Occasionally you'll need to make several parts precisely the same length. An old trick doing this is to make that dimension the *width* of a piece of styrene sheet. Then score and snap the parts from the precut piece, and the critical dimension will be constant. This is especially handy in bridge construction, where even minor inaccuracies will be obvious.

Scribing your own grooves

The variety of milled styrene siding and roofing sheets in the Evergreen product line makes it unlikely you'll often have to scribe your own grooves in plain sheet, but eventually you'll find a special project that requires doing so.

The scriber can be simple. A surplus dental pick, sharpened to the shape of the groove you want to cut, is a common scriber, and better hobby shops will carry a special tool made for the purpose. In a pinch — or if you have only a few grooves to make, the back of the tip of a No. 11 hobby knife blade can do the job.

The technique is a lot like scoring a sheet for the score-and-snap technique, except you want to remove material from the grooves. Hold the straightedge tightly against the sheet and draw the scriber toward you.

Start with a light cut to establish the line, then go back with more pressure to remove material. With just a little practice you'll be surprised at the uniformity of depth and width you can achieve simply by feel.

If the scriber raises a burr-like edge on one or both sides of the groove, scrape or sand it away. A fine grade (400-grit or higher) of wet-or-dry sandpaper, used wet, will quickly remove the burrs without marring the surface of the sheet.

Assembling sheet and strip parts

You can assemble fabricated styrene parts almost as fast as you can make them, since there's no waiting for the glue to dry. Just hold the parts in place for a few seconds until the solvent starts to evaporate. On small parts, the joint will set almost immediately.

Because Evergreen sheet and strip do not craze as readily as injection-molded kit parts, you don't have to worry about getting small amounts of solvent on the surfaces of the parts outside the joint. The styrene surface may turn slightly glossy where the solvent touched it, almost like a stain, but paint will make the stain disappear completely.

You do have a couple of seconds to move the parts or pull the joint apart if you did not position the parts correctly, but anything longer than that will result in a blemish on the surface of the parts – or a gooey mess. If the parts are simple, discard them and make new ones.

When you don't want the solvent to go somewhere, for example, when you are building a model that will come apart into several sections, the trick is to defeat the capillary action.

Solvent can't flow past a groove or trench, so scribe or file a groove where you want the bonding action to stop. A surprisingly small groove will do the trick. Chamfering the corner of an opening is another way of preventing solvent flow.

This charming country church in 1/64 scale is the work of Geoffrey Stippes. He built the model as a gift for his wife to commemorate their wedding — which took place in the full-size church!

Chapter 4

Scratchbuilding Basic Boxes

ASSEMBLING BOXES of various sizes and shapes is a fundamental building skill, particularly if you model trains, military vehicles, or structures. Styrene is ideal for boxlike assemblies because of its uniform strength and ease of both fabrication and assembly.

Let's start with layout

The first step in building anything from scratch is to lay out the parts. This can be as simple as drawing pencil lines directly on styrene sheet, or as complex as producing detailed plans and engineering drawings. Most modeling projects will fall somewhere in between, usually at the simpler end of the spectrum.

Jack Burgess' structure project on pages 22 and 23 employs templates made on a computer, and if you are proficient with such a program it greatly simplifies the layout chore. You can also use published drawings as templates by photocopying them to the scale of your model and taping or gluing them to sheet styrene. The sidebar on pages 20 and 21 provides the simple math you'll need to calculate enlargement and reduction percentages.

When you are designing a model, it's often worthwhile to first make a full-size mockup from cardboard. This extra step helps you think through how the model will go together, and lets you evaluate the shape and size in three dimensions. (It also means you can throw mistakes away and try again without wasting anything but cardboard!)

Just about any rigid, single-ply cardboard will do for this, including printed product packaging. If you cut the cardboard parts carefully, you can use them as templates for making the model parts later.

Scale rules and other measuring tools

My most essential layout and assembly tool is the right triangle. This can be a drafting triangle, a machinist's square, or even a homemade styrene item. I have an assortment of machinist's squares and prefer them over plastic tools because most solvents and glues won't stick to them. Their weight also makes them handy as assembly fixtures.

A 12" metal scale rule is indispensable for cutting styrene sheet, and you'll need a rule graduated in your modeling scale for measuring and laying out parts. Chapter 2 shows a

Geoffrey Stippes' Sn3 (1/64 scale) boxcars (above) are all styrene. Brian Ellerby made the S scale sheds at right from Evergreen sheet and strip using techniques explained in this chapter.

vernier caliper graduated in HO scale, which allows accurate measurements. Triangular architectural and engineering scales come in handy when your model corresponds to their graduations.

Dial or digital calipers can be used for scale measurements, and the new digital calipers are really easy to read. Simply convert the measurement on the real thing to inches and divide by your scale on a calculator. Here's a simple example:

4′-6″ = 54″ /48 (for 1/48 scale) = 1.25″

Use drafting dividers to lay out repeated spacing, such as for windows. If you use the points of the divider to make prick marks where you need to cut, you'll be able to feel the point of your knife contact the tiny indentation. This gives you a far more accurate measurement than a pencil line, which has a width of its own.

Drawing parts on plastic

To draw parts on plastic sheet, always use a metal straightedge and a square. Check right-angle corners on the sheets with a square, even if

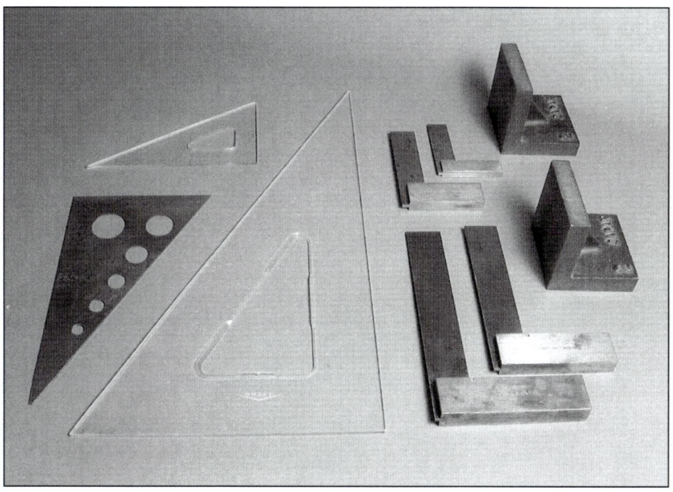

Crisp 90-degree angles are essential for building accurate models, and these are the tools to help you achieve them. From left to right: Metal and plastic drafting triangles, four metal machinists' squares, and a pair of heavy toolmakers' right-angle clamping plates.

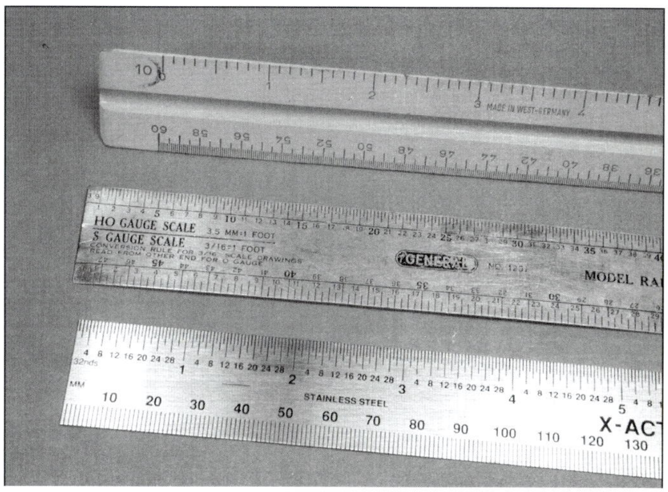

You'll need a stainless-steel scale ruler for your scale to measure and lay out scratchbuilt models, and to cut sheet styrene. The triangular engineer's scale, top, comes in handy when either your plans or your model correspond to one of its six graduated edges.

Scaling down prototype dimensions got a lot easier with the advent of digital calculators. They're inexpensive, so you can have one that stays on your workbench. The digital-readout calipers are equally handy for transferring dimensions.

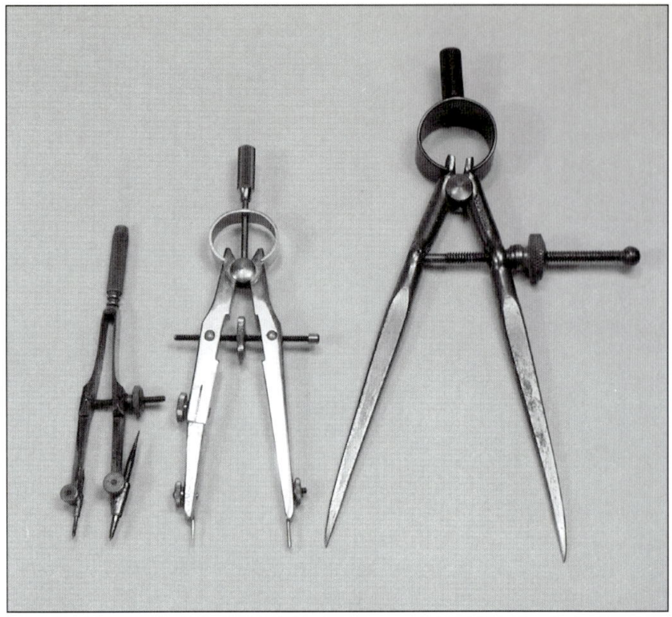

Dividers are handy for marking repetitive dimensions such as window spacings or rivet locations during layout. Make sure the divider legs end in sharp points so you can use them to prick accurate locations on the styrene.

SCALING UP AND DOWN

Ready for your first math lesson in years? Our challenge is to produce accurate scale drawings in the size of our model, so we can compare model parts to the drawing and use pieces of the drawing as templates.

What scale is your plan?

The first tool we'll need is a calculator, and the first calculation we need to make is the actual scale of the plans we have. (Just because they say "1/48 scale" down in the corner doesn't mean it's true.)

Start by finding a long, labeled measurement on the plan. Let's say it's 50 feet. Convert that to inches:

50′ x 12 = 600″

and divide the inch measurement by the stated scale of the drawing:

600″ ÷ 48 = 12.5″

Now measure the labeled measurement on the plan and compare it to the number you just came up with. Suppose it turns out to be 12.3″ instead of 12.5″. The plan is some scale other than 1/48!

We now know that 12.3″ on the plan equals 600″ on the real thing. If we divide 600″ by 12.3 we come up with 48.78 — so the plan is actually 1/48.78 scale. This is not at all unusual with published plans.

Calculating an enlargement or reduction percentage

Suppose we want to build a model in 1/35 scale. That means we'll have to enlarge the plans we have.

The formula is:

$$\frac{\textbf{Scale of drawing}}{\textbf{Desired scale}} = \textbf{Conversion \%}$$

We divide the scale of the drawing we have by the scale we want, and express the result as a percentage:

48.78 ÷ 35 = 1.39 = 139%

The same calculation will work if we are building in a scale smaller than the plans, say, 1/72 scale. We again divide the scale we have by the scale we want:

48.78 ÷ 72 = .68 = 68%

One easy way to check that you're doing the calculation correctly is to

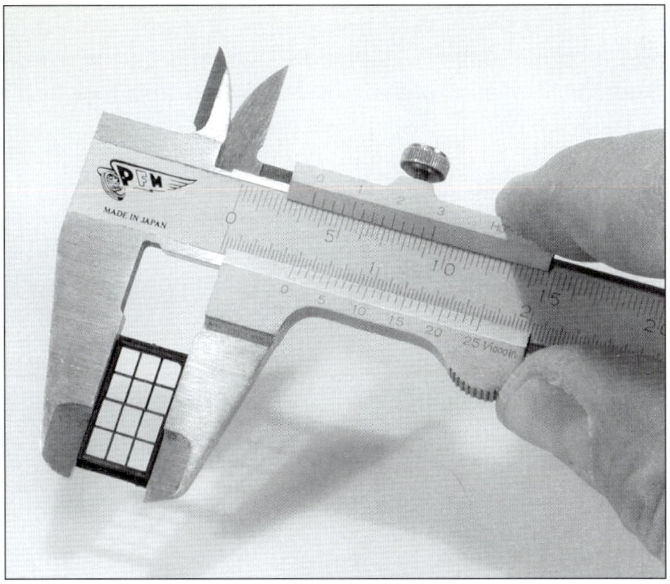

To accurately size window and door openings for commercial castings, measure the outside back edges of the casting with the jaws of the calipers, then transfer the dimension using the inside-reading blades on the back edge of the tool.

A sheet of 1/4″ glass makes an ideal flat assembly surface for boxlike models. The machinists' squares and toolmakers' blocks are used to check alignment and to prop the parts in place during assembly.

remember that if you're building in a scale larger than the plans you have, the percentage must be greater than 100; if you're going smaller, the percentage has to be less than 100.

To the copier!

All sorts of retail outlets have reducing and enlarging copiers, and the best kind to use is one where you get to make the copies yourself. (If you have a computer scanner and software to enlarge or reduce scanned images, the procedure is essentially the same.)

Take your scale rule with you to the store, so you can check that the copier is actually making the reduction or enlargement you specify — copiers can be off.

Start with a test copy, check a labeled measurement with your scale rule, and adjust as necessary. Once you have the right settings, it's a good idea to make at least two sets of plans in your construction scale: one to keep as reference, and the second to cut up for parts templates.

It's legal to copy published plans for your own use, and most modeling magazines and books include notices to that effect.

Common scale conversions

Here's a handful of conversion percentages between common modeling scales. Remember to always check the drawings you have to verify they are the stated scale before enlarging or reducing them.

Drawing scale	Model scale	Percent
1/16	1/24	67
1/24	1/25	96
1/24	1/48	50
1/25	1/24	104
1/32	1/48	67
1/32	1/35	91
1/35	1/48	73
1/35	1/72	49
1/35	1/32	109
1/48	1/32	150
1/48	1/35	137
1/48	1/72	67
1/48	1/64	75
1/48	1/87	55
1/64	1/48	134
1/64	1/87	74
1/87	1/48	181
1/87	1/160	54
1/96	1/192	50
1/144	1/192	75
1/160	1/87	184

you are using the edge of a brand-new sheet as your finished edge. Armor modeler Joe Morgan adds, "Although all the Evergreen sheets I've ever used have been square, it's simply good modelbuilding practice to check – every time."

For boxlike models, lay out all the sides adjacent to one another on a single sheet of styrene so the height is identical. This is where the dividers come in handy: Use them to verify that all the walls are the same height, and that opposite walls are the same length. The old adage about measuring twice, then cutting once is still great advice.

The thickness of the styrene that you use to make the box must be taken into account when laying out the model. For instance, if you use .060″ styrene for the walls of a rectangular structure, either the ends or the sides must be shorter by .120″ to compensate for two thicknesses of the overlapping material.

For models with peaked roofs, it's generally better for the gabled end to overlap the sides, so the roof material can rest on the ends and tops of the sides.

If you are building a structure and will use commercial window and door castings, a vernier or digital caliper makes laying out openings for them easy. Clamp the jaws of the caliper lightly on the back of the casting (the part that fits in the wall opening), then use the blades of the caliper (the part used to take internal measurements) to transfer the exact dimension to the correct location on the model.

Cut and assemble

When you've completed laying out the model and checked the dimensions twice, scribe and snap the parts as described in Chapter 3. If the box has more than four walls, use masking tape to assemble them temporarily to see how things fit. Label the pieces as you cut them to help you keep things straight during assembly – once you've glued together a few pieces upside-down, a note like "top" doesn't seem so silly.

Thin styrene wall sections (.010" .015", or .020") will need bracing to hold them flat and rigid. This is especially true of scribed sheet, which is likely to bow. You can use strip stock

ASSEMBLING ANGELO'S

by Jack Burgess

Angelo's is a typical scratchbuilt styrene structure that illustrates the steps in fabricating and assembling such a model. It's HO scale (1/87), but the size is immaterial to our discussion.

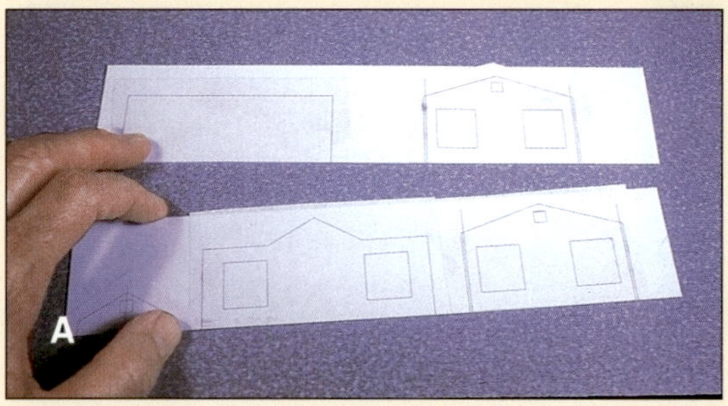

A

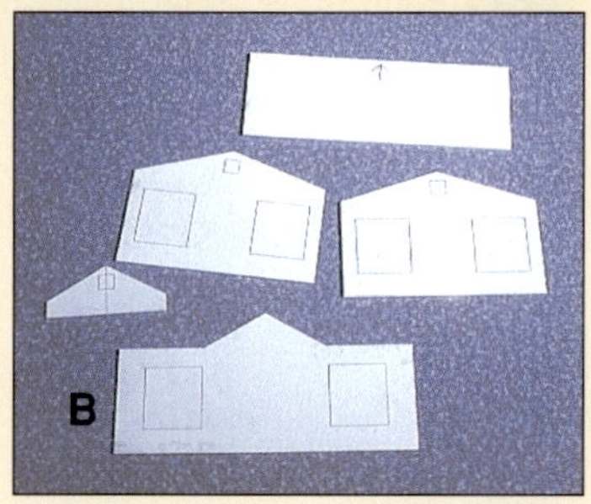

B

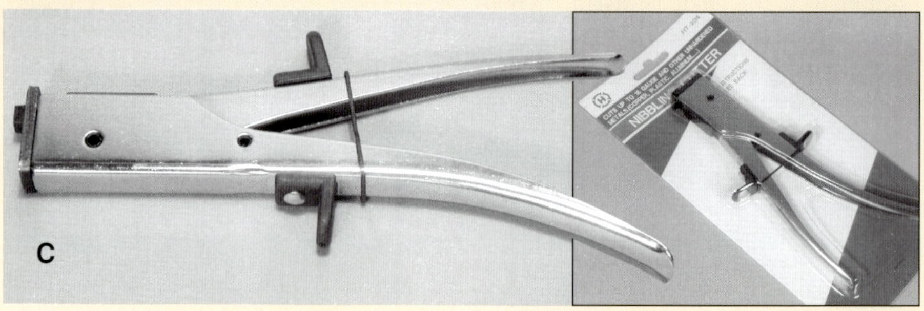

C

I drew the building using CAD (computer-aided drafting). My plan was based on a just one available photo and a number of what I considered reasonable assumptions. Once the building was drawn up, I printed out cutting templates for each of the sides with the required openings for the windows and doors outlined. These templates were temporarily attached to Evergreen milled styrene siding with a low-tack spray glue used for holding stencils in place, **A**. If you have printed plans, you can photocopy them to scale and glue the copies to the styrene.

The next step was to cut out the sides using the templates as guides, **B**. Cutting window, door, and similar openings can be time consuming. The traditional approach is to deeply scribe the outline of the opening, then scribe an "X" from inside corner to inside corner. Then, if you are lucky, you can snap out the inside pieces. However, you still normally have to file the corners square.

Using a nibbler, **C**, is both faster and easier. Nibblers or nibbling tools are used in the electronics industry to cut large holes in sheet metal and are available from Radio Shack and hobby tool dealers.

Make a starter hole near the center of each opening. While you can use a drill bit, it is just as fast to hold an X-acto knife with a No. 11 blade vertically and simply twist it to create a hole. Enlarge the hole by carving with the blade until the head of the nibbler can be inserted.

You then "nibble" up to the outline of each opening, **D**. Since the head of the nibbler is square, it is easy to work up to a corner, turn 90 degrees, and nibble down to the next one. If you are careful, little or no cleanup will be needed and the opening will be exactly to size.

One side and one end are being assembled on a flat surface, **E**. A couple of machinist's 90-degree clamping blocks are used to hold the two sides plumb, while the orange draftsman's triangle ensures that the joint is square and the edges of the two pieces are flush. The other side and end are assembled in

for bracing, or cut backing ribs from .040″ sheet where they will not show.

Make window openings by scribing around and diagonally across the opening, then snapping. Jack Burgess recommends a nibbler, below.

Flat and square construction

You'll need a flat surface to assemble models. A sheet of 1/4″-thick glass works well, and a glass shop will gladly cut a piece to size and bevel or polish the edges for you. (Most older modeling books call for plate glass, but modern glassmaking techniques yield a flat surface even on common glass.)

Micro-Mark sells a wide variety of machinists' squares and toolmaker's right-angle plates. I consider them essential tools for building square and plumb assemblies. In addition to checking assemblies for squareness, you can often use the squares to hold parts in position while bonding them. The drawing (page 26) shows a simple trick to ensure that solvent does-

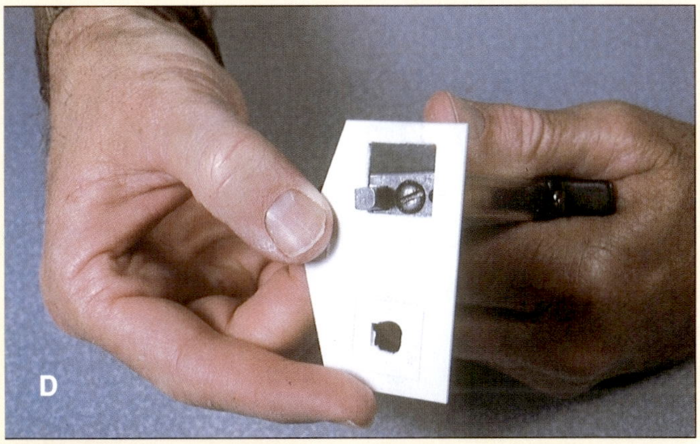

D

E

F

G

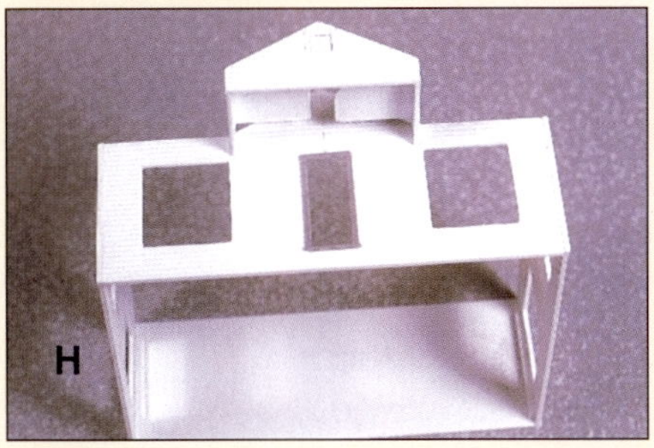

H

I

J

a like manner and then the two subassemblies glued together to form the basic building shell, **F**.

I glued square strip styrene into the corners to reinforce them, **G**, then assembled the porch overhang and added it to the building, **H**. The porch posts were cut long, inserted with the extra length hidden up inside the porch overhang, and glued in place. This was much easier (and stronger) than trying to cut the poles to exact length.

The roof pieces, **I**, are plain Evergreen sheet, cemented from inside. I cut the rafter tails, **J**, over-length, cemented them in place, and trimmed each one even with the roof using flush cutting tweezers. I did this with the building resting upside down to ensure the rafter tails were cut perpendicular to the floor.

At this point the building shell was complete, ready for painting and adding shingles and signs.

THE NEB&W SNAP METHOD FOR SCRATCHBUILDING BRICK STRUCTURES *by John Nehrich*

One of the best-known model railroads in the U.S., if not the world, is the HO scale New England, Berkshire & Western. Located on the campus of Rensselaer Polytechnic Institute in Troy, New York, the NEBW re-creates much of the downtown of Troy in the steam era. That means building lots of brick buildings from scratch. Here's how John Nehrich tackles the job.

We make a mockup in cardboard of just about every structure before we start building. This is perhaps the most important and easily overlooked step. Often we wind up making a second or even a third mockup, to ensure the building fits into the overall scene.

Brick sheet, styrene, and windows

The best HO brick material is Holgate & Reynolds vinyl sheet. It comes in large sheets, is thin enough to be easily worked, and has the closest-to-scale-size bricks. It is so thin, however, that it needs substantial backing. At one point I had standardized on .040″ styrene sheet, but have since come to prefer .060″. Its thickness requires less bracing than .040″, yet it isn't that much thicker to make it hard to work.

I build sub-shell walls that have all the openings cut out, glue the H&R brick veneer over this, then cut through the brick at each opening. The whole method revolves around the fact that masonry structures rarely have window frames on the surface of the wall. The wall is thicker, so the window is set into the wall, and the frame would serve no purpose.

There are few HO castings available for frameless windows, so we have to make do with window castings with frames. Rather than cut off the frame, the window can be inserted from inside the wall. To maintain the proper relationship of upper and lower sash, the casting must be turned upside down. Most of the Grandt Line castings can be used in this fashion, as they are detailed front and back. Few castings from other manufacturers are suitable.

Snapping and welding

Since the edges of the openings will be visible, it is important they be sharp and straight, unlike the wavy cuts that the window frames would hide on a wood structure. Cutting an interior hole in any material is difficult. However, styrene has two important advantages over other materials: It snaps cleanly along a scribed line, so you don't have to cut all the way through, and it can be welded back together using solvent cement.

We make the openings by snapping each wall into horizontal bands, **A**, discarding the pieces where the openings will be, and welding the wall back together, **B**. Once the wall is dry (overnight), it is sanded with coarse sandpaper to remove ridges raised by the scribing-and-snap process, and to make the entire face in the same plane in case of any misalignment when gluing. I sand both sides.

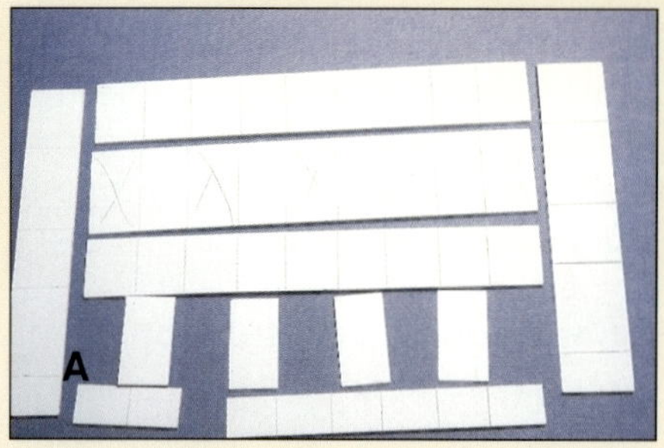

Applying the brick sheet

Weldbond water-soluble contact adhesive is the glue I use to attach the brick to the sub shell. It's water-soluble, so there's no solvent to get trapped between the brick and the sub shell, and as a contact cement it does a good job of attaching the vinyl brick to the nonporous styrene.

The instructions say to coat both surfaces, let them dry, then position. No way! I coat both surfaces, scrape off the excess using a scrap of styrene, and bring the two pieces together while the adhesive is still wet. This allows for shifts in alignment. I apply a little super glue at the corners to keep the materials aligned overnight. (The brick could be attached with super glue alone if you can spread it evenly before it sets up. This is useful if you don't want to wait overnight for the next step.)

I cut away the brick material using the sub shell as a cutting guide, **C**, and finish each opening with a file. I also lightly sand the surface of the brick at this stage to give a matte surface. Bracing is added with Evergreen .060″ by .250″ strips, **D**.

The corners could be mitered, but I just make the brick material on one wall extend past and cover the edge of the adjacent wall, **E**. You have to take this into account when you are laying out the walls as to which wall will be shorter by

n't seep onto your assembly surface and ruin a finished model surface.

Large wall pieces can be cumbersome to hold in alignment while bonding. Here the trick is to bond a stop or corner brace along the edge, spaced the exact thickness of the mating wall from the edge. The drawing (page 26) shows an easy-to-make styrene fixture to accomplish this. With the stop in place, it's easy to make an accurate corner joint, even if

D

E

F

G

the thickness of the brick covering. I use Walthers Goo to attach the brick to the edge of the styrene wall at the corners, followed by flowing super glue into the joint, using a piece of wire as an applicator. Subroofs are plain styrene sheet, **F**.

Painting — now or later?

At this point you could paint the wall assembly and all the windows, sills and other trim, if they are to be a separate color. I used to do that.

Nowadays I complete assembly before painting unless I expect a real problem in getting a brush into an area. This makes the building look more like its prototype and more like the finished product, so I get more motivated to keep working on it. Second, I don't feel I have to be 100 percent satisfied with the treatment of the wall before I go any further. Since I'll be painting the walls with the castings in place, I can always repaint them.

When it does come time to paint, I usually paint the whole building with Floquil railroad color, brushing it on the styrene and vinyl surfaces without benefit of Barrier. Even if I plan on painting the trim different colors, I paint everything a base color first.

Arches, lintels and sills

To make brick arches I cut a strip , one brick wide from the edge of the Holgate & Reynolds sheet. By cutting part of the way between each brick, the strip curves, like flex track. This is glued in place with a little Goo, followed by super glue applied with a wire to flow under the arch. In a few cases I've laid out the arches on .015" styrene with a compass and scribed individual bricks in the styrene.

For stone lintels over the windows, I use .015″ x .080″ or .015″ x .100″ styrene strip. For the sills under the windows, I use scale 2x4 or 2x6 strips, **G**. I've found that just letting solvent flow under the styrene dissolves it enough to stick to the vinyl, even though the vinyl brick surface itself doesn't dissolve.

you only have two hands!

When you've assembled the sides and ends of your box, it's time to add a top and bottom (or roof and floor). Rather than cutting them to dimensions taken from your plans, make these parts to fit the model. That way, minor inaccuracies in the sides and ends won't be amplified in the new parts.

Removable structure roofs are easy to make with styrene. Add strips or

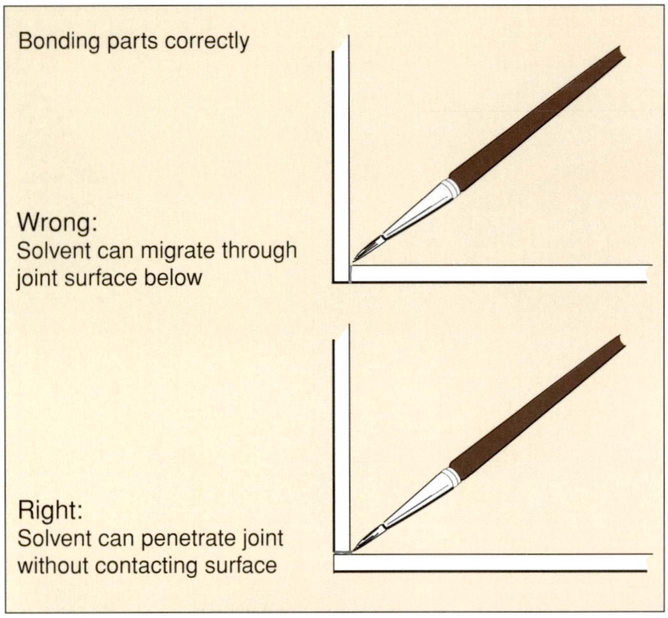

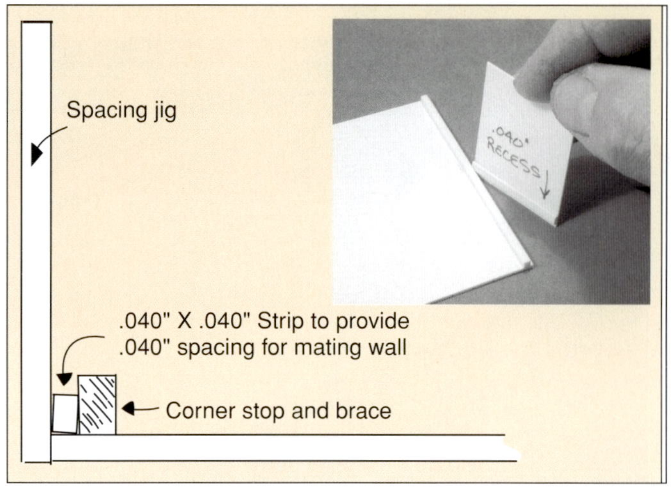

Here's a simple way to provide a recessed alignment tab along one edge of a model wall. The gauge, made of scrap styrene, ensures the tab will be the same depth as the thickness of the mating wall.

Molded styrene windows and doors come in all scales, and all shapes and sizes. Particularly in the smaller scales they are far easier than fabricating your own parts.

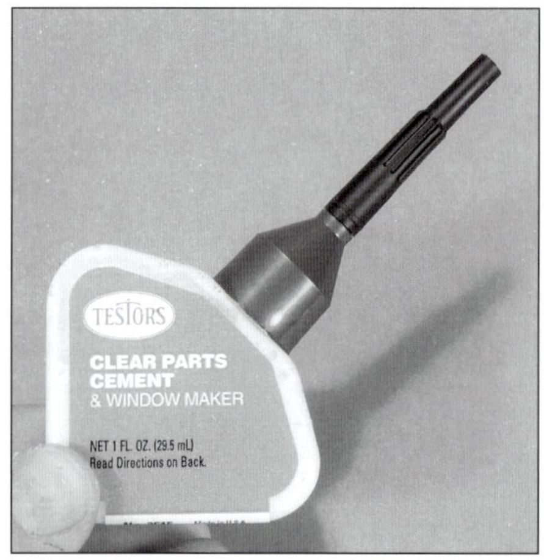

Cementing clear parts is tricky, since glues will craze or discolor them. Testor makes a special cement, or you can use a white PVA glue such as Elmer's.

gussets to the bottom of the roof that are a slip-fit with the walls, then be careful not to bond them to the walls during subsequent assembly.

Dozens and dozens of molded styrene door, window, and detail castings are available in the most common modeling scales, and they greatly simplify structure modeling.

Test-fit each one in its opening before assembly, and be particularly careful to align the bottom edges of windows. Wherever possible, bond the parts by applying small amounts of solvent from the back of the wall.

Working with clear styrene

Clear styrene, whether sheet stock or injection-molded, is more brittle than white or colored. This makes it slightly more difficult to scribe and snap clear sheet, and you'll find you have to cut most of the way through it before snapping.

The main use for clear plastic is as window glazing, where you want to maintain crystal clarity. This calls for care in assembly, because even tiny amounts of solvent will etch or craze the surface, turning the window translucent.

Substituting super glue for solvent isn't the answer either, because it causes a frosty white residue to form on clear styrene parts. Try white glue such as Elmer's (polyvinyl acetate, or PVA). Testors offers a water-based Clear Parts Cement (No. 3515) specifically for this application.

Pat Hawkey built this handsome 1/48th scale TBF-1 Avenger from an Accurate Miniatures kit. The WWII torpedo bomber model owes most of its fineness to a modern, highly detailed kit, but Pat made sheet styrene open cowling gills to add an extra bit of life to an otherwise static model.

Chapter 5

Working with Styrene Kits

MOST MODEL KITS are made from styrene, and the techniques used to assemble them are similar to the techniques for assembling Evergreen sheet and strip parts. In this chapter we'll explore those techniques, and begin to show how Evergreen materials can be used to augment and improve kits.

Sizing up the job

The first step in building a kit to always — ALWAYS! — sort out the parts to see what you've got. At the very least, you should inventory the parts to determine whether anything is missing. This is also the time to look over the instructions from beginning to end to see how the parts go together.

Most modelers will admit to a philosophy of "When all else fails, read the instructions." This can and will get you into trouble. Kits are becoming more and more sophisticated, particularly in the way they hide joints. This means the way the parts fit together may not be obvious.

As you look through the instructions – nowadays, they're mostly drawings – you'll want to think about breaking the model down by subassemblies for painting. Kit manufacturers don't always concern them-

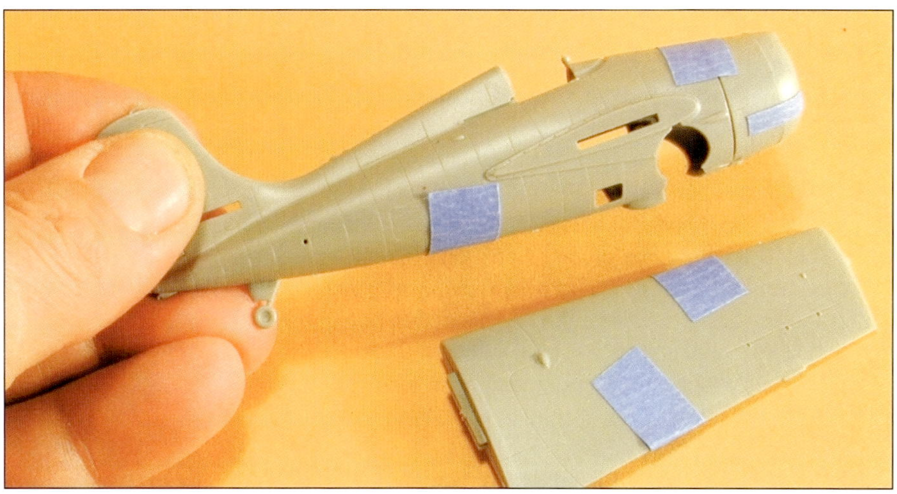

The fit of the fuselage and wing parts on this Hasegawa 1/72 scale F4F-4 Wildcat is nothing short of spectacular, but it still makes sense to dry-fit the parts with masking tape to see how the model goes together.

The large green hull molding on this old Aurora schooner kit wasn't quite flat along the joint line, and the aligning pins hindered more than helped. Here a sanding block is being used to flatten the mating surface and eliminate the pins to make more accurate fitting possible.

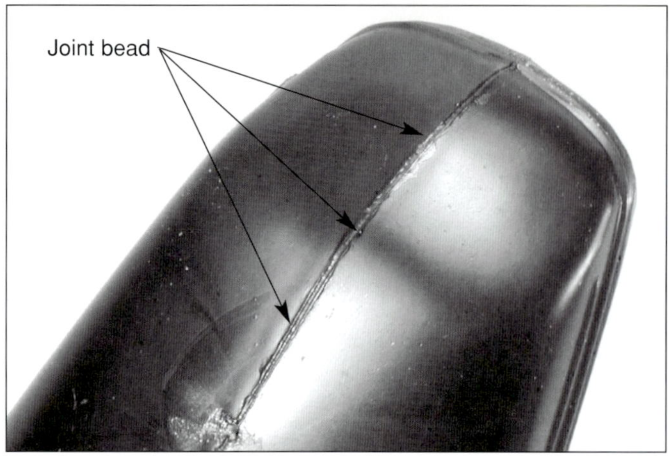

By softening both mating surfaces of this ship hull with solvent, then joining them with a little extra pressure, this small bead of molten styrene was squeezed out of the joint. When the bead hardens, the joint can be sanded and polished smooth with little or no filling required.

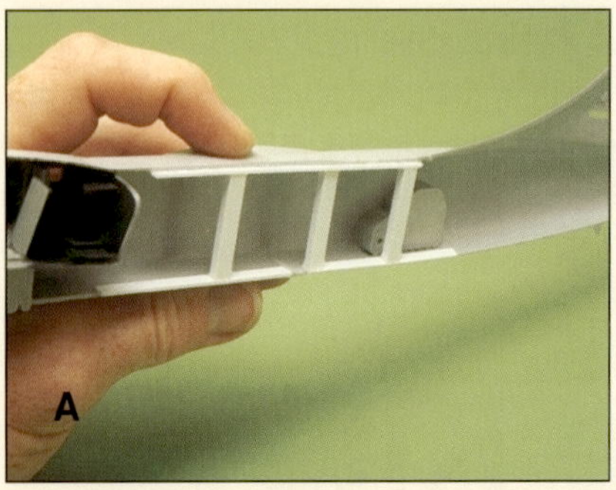

REINFORCING KIT JOINTS AND SEAMS

by Pat Hawkey

I often use styrene sheet and strip to strengthen internal model joints and provide more gluing surface. Especially on large aircraft models, fuselage halves and wing leading edges offer relatively small bonding surfaces. With all the handling the model will receive during construction and finishing, it's easy for these joints to fail — and once the model is painted, a split fuselage seam is a hateful problem to deal with. Therefore I make a point of building extra strength into any doubtful area.

This Italeri B-66 features a long fuselage unbroken by windows, turrets, or similar openings. To remove the flex potential, I fabricated an inner box of long strips top and bottom braced by three crosspieces, **A**. This provided plenty of gluing surface fore and aft, and the vertical braces eliminated any chance of flexing.

This Italeri V-22 also has a long unbroken fuselage line, and the moldings were slightly warped to boot. With the main floor fixed solidly in place, I added a wide strip of .040" strip stock and forced out the dip in the belly with a crossbar, **B**.

I believe plastic airplanes simply can't be built too strong. This Monogram 1/48 scale Me 262 shows a pair of mating surfaces added to ensure a solid joint between fuselage and wings, **C**. This photo also shows another useful role for sheet stock: filling landing gear openings. If you build aircraft in the in-flight configuration, you'll soon discover most kits weren't meant to be presented that way. As often as not, the fit of gear doors in the closed position is something of a joke. Especially if there are multiple doors to fit into a given opening, it's easier to replace the kit parts with a single piece of sheet cut and sanded to fit.

Here are two more examples of the same situation, **D**. Hasegawa 1/48th P-51 on the left and Monogram 1/48th A-37 on the right.

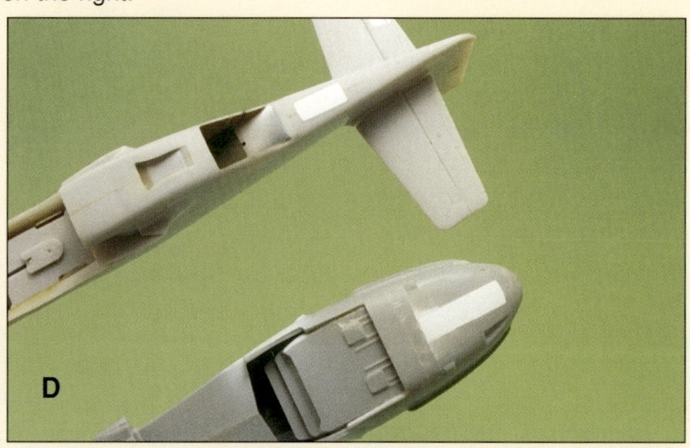

selves with this, but it's a vital aspect of building top-notch models.

Make notes on the instruction sheet showing the places where you plan to pause in assembly to paint or finish a subassembly. You may change your plan as you actually work with the model, but at least you'll have one.

To clean or not to clean?

A lot of experienced modelers recommend washing injection-molded kit parts before detaching them from the sprues. This removes mold-release lubricants that can prevent solvents and paints from sticking. I don't perform this step, and I've never had a model where the solvent wouldn't bond the parts. When we talk about finishing, however, I'll recommend a thorough cleaning of the assembled model before applying any paint or primer.

To fill a long, thin seam such as this sawdust-burner joint, place masking tape along both sides to protect the adjacent detailed surfaces, then apply filler and sand. Nitro-Stan 9001 is a common fine-grain auto body putty, and a toothpick is a good way to apply it. From the looks of the battered tube, it's seen plenty lot of use.

Removing flash

Plastic is forced into steel molds under extremely high pressure to make injection-molded parts. Where the halves of the mold come together they leave a thin line called a parting line. Some manufacturers are clever enough to make these correspond to raised detail lines on the real thing, but usually you'll have to remove them by scraping, sanding, or both.

When molten plastic seeps between the parts of the mold, it forms thin, featherlike projections called "flash." Old kits – molds that have been in production for decades – often have this problem. Flash is something that shouldn't be there; remove it by trimming with a brand-new No. 11 blade and by sanding the area smooth where required.

Looking for trouble

After removing the parts from their sprues and cleaning up the attachment points, begin assembly by dry-fitting the major kit parts. Masking tape makes a convenient "third hand" to hold parts together temporarily, and you may want to use it during final assembly, too.

The little pins and holes provided on kits to help align the parts aren't always helpful, and occasionally you'll want to remove one or more of them to allow you to adjust the fit. You'll also find problems where the mating parts aren't perfectly flat along their joint; the solution here is to sand the edges smooth for a better fit.

Joints between major parts require extra care. Not only are these areas often prominent on the finished model, but they also need to be extra-strong to hold up during subsequent construction and finishing steps. If an aircraft fuselage or ship hull splits apart as you prepare to paint the model, there's usually not much you can do about it except buy another kit and start over.

To strengthen these joints, start by applying a slow-acting solvent to both edges to soften them, then bring the parts together and apply more solvent from the inside of the model. Press the parts together until a small bead of softened plastic oozes out along the joint line. When this bead hardens, you can often sand it smooth and make the joint disappear

The circular ejector-pin marks on this plastic trestle-bent molding will be obvious on the finished model. The mark on the right side has been filled with a drop of thick super glue, then quickly set with accelerator. The next step will be to sand it flush.

FILLING GAPS WITH STYRENE SHIMS

by Pat Hawkey

Perhaps the most basic use the aircraft modeler has for sheet styrene is that of the gap-filling shim. While the mold-making technology of the 1990s has resulted in kits whose fit is nothing short of phenomenal, you don't have to go too far back to find models whose fit leaves much to be desired.

Significant gaps and cracks (not the hairline variety) are easily and effectively eliminated with an appropriate width and length of styrene. The advantage of filling a gap with styrene instead of thick cyanoacrylate (super glue) is that the styrene is the same density as the surrounding model and therefore easier to trim, sand, and if need be, polish. Super glue cures much harder than the styrene around it, and the filing and sanding needed to remove it from where it's not wanted often results in damage and loss of surface detail.

Here the subject is a Hobbycraft F7U Cutlass. The job was to "re-hab" the model, which had been previously built and painted — not very well, and not by me — and I decided to strip and rebuild it. There was a significant gap where the tail met the landing-gear fairing, **A**.

The fix was to insert a bit of .020" sheet stock, secured with a couple dabs of liquid cement, **B**. When the cement was dry, I trimmed the excess with a pair of sharp manicure scissors, **C**.

Next I went in with a piece of emery board to rough sand the area smooth, **D**. Not only was there a gap to fill, but the contours of the joined areas didn't match very well either. One way to fix this is to keep on sanding until everything is flat, but that method will result in an inaccurate shape to the model.

This is where gap-filling super glue, **E**, is employed. The advantage to the super glue is the speed with which it dries. I filled the depressed area with super glue, and a touch of accelerator cured it nearly instantly.

Now the area is sanded smooth again, any scratches or slight depressions being filled with body putty, sanded again, primed, and wet sanded one last time with very fine grit wet-or-dry sandpaper, **F**.

A

B

C

D

E

F

without using any filler at all.

Bear in mind that you don't always have to cement the entire joint at once. Where possible, reinforce major joints with strip or sheet styrene (see the sidebar, page 28). Once the solvent has dried, also consider reinforcing the inside of the joint with a fillet of thick super glue or 5-minute epoxy.

Correcting warped parts

Minor warping on thin parts can be worked out by gently bending the parts during assembly and holding them straight while the solvent sets. Thick, badly warped parts can require softening in 180-degree F. water (not boiling!) and straightening against a flat surface such as a sheet of glass. Where this won't work, the solution is to make multiple thin slots with a razor saw on the back of the part to render it flexible enough to bend it into place, then fill the slots where they are visible.

The Flexi-File is a U-shaped frame with a thin band of Mylar-backed wet-or-dry sandpaper stretched across the open end. It's an ideal tool for smoothing joints on round parts such as aircraft fuselages or this smokestack.

Choosing and using fillers

Kit manufacturers divide a model into parts to mold it, and the joints between the parts may be visible when the parts are assembled. When they shouldn't be, we need to disguise them – usually with a filler.

There are many commercial fillers available, and you should experiment until you find one that you like. The old standby is auto-body filler, used to smooth out imperfections in – you guessed it – automobiles, before painting. Several of these are repackaged and sold in hobby shops as filler putty, and since it's unlikely that you'll ever use up two pounds of fine-grained auto body filler, that's the kind to buy.

The main problems with one-part auto body fillers stem from the fact that they dry by evaporation. This means that drying takes time – often overnight – and that the putty may shrink, making another application necessary.

Filler putties are often quite porous, and on models this can show up as pinholes in the painted surface. On full-size automobiles the surface imperfections are sealed with a heavy-bodied primer, and the same approach works for models, especially large-scale ones.

Another way to seal the putty is to smooth on a thin coat of super glue, then wet sand. This also reinforces the surface of the putty during subsequent finishing operations.

Filling gaps with super glue

An excellent alternative to filler putty is gap-filling super glue. This thick super glue stays where you put it, including in a gap that you want to fill. A quick spritz of super-glue accelerator will make it set instantly, which means that you can proceed with sanding, more filling, or painting without waiting for the filler to dry.

There's an important caution when using super glue as a filler: Don't fill a joint and put it aside before filing or sanding the filler smooth. When freshly set, super glue has roughly the same hardness and workability as styrene, but in 24 hours it turns rock-hard. This is good for the model in terms of durability, but if you wait too long to smooth the filled joint, you'll have to work a lot harder to do it.

Gap-filling super glue is ideal for eliminating the circular ejection-pin marks you'll occasionally find on the outer surfaces of kit parts. Just apply a drop of the glue to the depression, hit it with accelerator, and sand flush.

Sanding joints and fillets

Smooth either putty or super glue joint fillets with files and sandpaper. If you've never tried wet-sanding with wet-or-dry sandpaper, you're in for a treat. The water serves as a lubricant and flushing medium to remove sanded material from the working area, and this helps the paper remove material quickly and leave a smooth surface.

Start wet sanding with 400-grit paper, then go to 600-grit for a ready-to-paint finish. You can even remove scratches and scuff marks from clear parts and restore a crystal-clear surface if you wet-sand with polishing grits in the range of 1600- to 2400-grit.

Scrub out sanding dust with an old

toothbrush and detergent and water before proceeding with assembly.

Painting as you go

Where visible parts of a model will become inaccessible to paint after major parts are joined, you'll have to pre-paint them. This happens most often with aircraft fuselages and ship hulls.

After painting such areas and assembling the painted parts, protect them with tissue stuffed into the openings or mask them with tape and the plastic from food-storage or sandwich bags.

Adding clear parts

Clear parts and the adhesives for them were discussed in Chapter 4. Regardless of what the kit instructions say or show, whenever possible parts such as canopies and windows should be applied after all painting, decaling, and weathering operations are completed.

When the design of the model makes installing clear parts earlier a must, you'll have to mask or otherwise protect them from damage during subsequent steps.

When you've become proficient in kit assembly, it won't be long before you'll want to model something that isn't offered as a kit. One way to accomplish that is to adapt parts from two or more kits to make something distinctively your own. That's the subject of our next chapter.

USING AUTO-BODY FILLERS

by David Merriman

I don't use the air-dry filler putties sold in hobby shops, because I have yet to find a suitable one. They dry too slowly, shrink too much, and some just don't have the chemical compatibility to adhere to the many different materials I use for model fabrication.

Head to the automotive refinishing supply store and get some lacquer-based touch-up putty designed for automotive body work. I like Nitro-Stan 9002, available in either red or white. I find I have much better control of the putty if I lay it down on the model with a stiff brush.

While at the auto store pick up a can of Evercoat Euro-Soft or Spot-Lite isothermic-curing body putty. These self-curing two-part putties are a must when you're faced with a significant gap filling or part re-contouring chore. (Even the smallest can of Euro-Soft or Spot-Lite will last the average modeler a good deal more than one lifetime!)

This 1/35 scale Marine Corps M2A4 tank is the result of a complex kitbashing project that started with a Tamiya M3A1 kit. Joe Morgan built the model and took the picture.

SOMETIMES IT SEEMS as if there must be a kit offered for every subject you could ever want to build — and in every scale! It doesn't take long, though, to find something that isn't available. Often, the best way to model the subject is to modify a kit that's close, or use parts from several kits.

Some modelers call models built using this approach "conversions," while others call them "kitbashed." Whatever term you choose, modifying kits combines the best of what kits bring to the table — accurate forms and wonderful surface detail — with your specific modeling needs.

Chapter 6

Kitbashing and Conversions

In one sense, kitbashing or converting is really scratchbuilding, with kit parts as the "scratch." You can start either with a kit that you like and change its configuration to better suit what you need, or you can settle on what you want to build, then go looking for the kits than lend themselves to building it. Well-known model railroad structure expert Art Curren once explained the essence of kitbashing as "Seeing kits as parts."

Concept or technique?

That quote leads to a chicken-and-egg question: What's more important, the ability to see the a kit as grist for the model you want to build, or the techniques for building it?

I'd say strategy is more important than tactics here. The concept is the most important ingredient, since the techniques for kitbashing are mostly the same ones you use for building straight kits or for working with styrene sheet and strip.

So how do you learn to size up a kit in terms of what else you might build from it? One good way is to take a conversion project from a book or magazine and build it just the way the author did. By following the hands-on instructions, you'll quickly learn how the builder thought his way through the project.

For models that will be kitbashed from flat parts like model railroad structures, you can array the kit pieces on a copy machine and make paper parts to test cut and fit. Gluing the paper parts to cardboard will allow you to mock up the model in three dimensions.

For many conversions, the way the real thing was built can lead the way for you. Many prototypes — aircraft, automobiles, tanks, railroad equipment — went through documented

Revell's 1/177 scale WWII fleet submarine has appeared in a variety of boxes, including "USS *Lionfish*." David Merriman used it to model USS *Rasher*, a radar picket submarine. He lengthened the hull by grafting on parts from a second kit, and made a new sail from wood, fiberglass, and vacuum-formed styrene. The distinctive planked wood deck was represented with specially scribed pieces of .015″ styrene sheet.

Bob Steinbrunn converted Monogram's 1/48th scale Lockheed P-80C into a YP-80A (the service-test model). He used parts from two kits, extensively modifiying and detailed them with Evergreen sheet, strip, and rod.

evolutionary development, and kits of the versions before and after the one you want to model will provide the shapes and details you need.

When the real thing was itself a conversion, such as a two-seat trainer version of a single-seat fighter plane, you can often start with the standard version — or more likely, two kits of the standard version.

Modelers interested in World War Two German military vehicles need only follow the lead of the Wehrmacht, who became the all-time champs of converting captured and outdated equipment to new uses. Joe Morgan's Bison project in this chapter is a perfect example.

Conversion techniques

Only a few modeling techniques are truly specific to converting kits. Most involve cutting kit parts and bonding them together in their new configurations.

Spend plenty of time brainstorming with the kit parts before you make any cuts. Juxtapose the pieces to evaluate how they'll fit together and where the joints will fall. Often, you can position joints where they'll be hidden later by details, which means you don't have to finish them carefully. Although kitbashing is by definition a patchwork process, we don't want a patchwork result.

Try to work out the smallest number of pieces required to make the model. This will make your finished product stronger, and usually the pieces will be easier to align.

And finally, don't hog-tie yourself by trying to make the model entirely from kit parts. Especially when the model has flat surfaces, it can be easier to substitute styrene sheet, detailing the surface with surface features removed from the molded pieces.

Saws for styrene

After deciding how your conversion will be pieced together — and perhaps making a rough sketch — it's time to put saw to plastic. Because molded parts vary in thickness, the score-and-snap technique we use for cutting sheet styrene usually won't work. Instead, it's time to reach for the saws.

The most common type of model-building saw is a fine-toothed razor saw. It has a thin blade stiffened by a heavier metal channel along the back edge. As explained in Chapter 2, reversing the blade so it cuts on the pull stroke improves your control over the sawing action. (A larger brother of the razor saw called a dovetail saw or tenon saw may be just what you need for kitbashing larger models.)

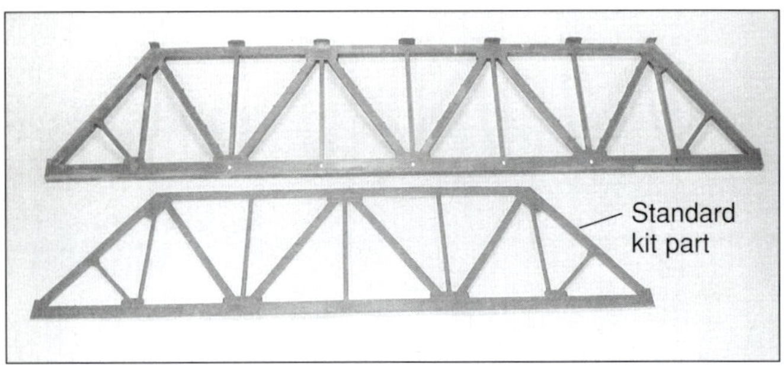

Bridges are easy to kitbash, because real bridges consist of repeating standard pieces. Here's a typical project: Parts from two Walthers HO (1/87 scale) kits were combined to make a bridge two bays longer than the kit. If the bridge were significantly longer the trusses would have to be higher for strength, but this conversion has good proportions and is convincing.

The thin-bladed jeweler's saw or its larger version, the coping saw, is handy because the blade can be removed from the frame and passed through a hole in the model to cut inside openings.

Very heavy plastic sections call for a fine-toothed hacksaw. Start with a brand new blade, and earmark it for use only on styrene so it will stay sharp.

Making the cut

The key to cutting plastic quickly is to work slowly. Start the saw with very light strokes, and once the cut is started, continue with light pressure. Let the tool do the work: If the saw does not cut quickly or cleanly, a sharper blade — not more speed or pressure — is what's needed.

When you can, use a joint line, panel line, or some other molded-in detail of the model as a guide for sawing, as well as for alignment during assembly.

Masking tape is excellent for laying out cuts and protecting parts while you cut. Plan to cut parts wide of the joint, then file, sand, and trim them for a precise fit.

It's common for the saw blade to jam when cutting thick sections; if this happens, lubricate the blade with Vaseline or vegetable oil, then remove the residue with dishwashing detergent after sawing.

When the major parts have been

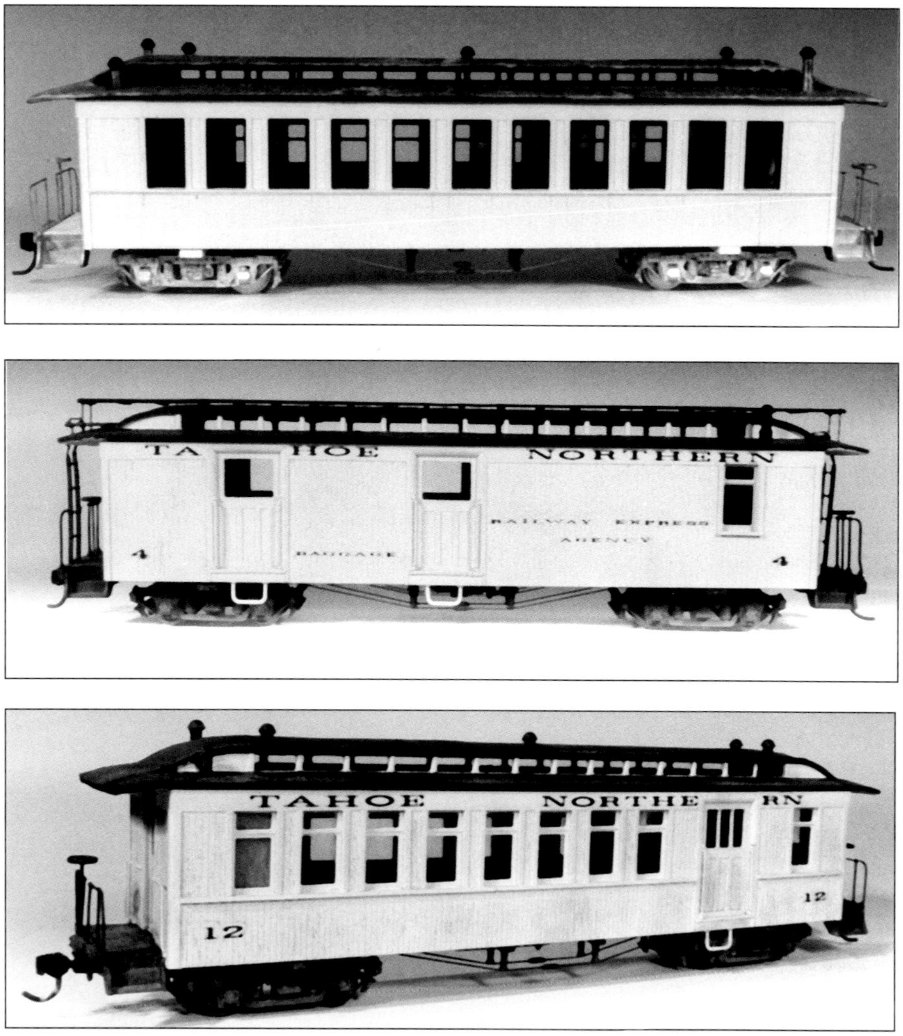

An excellent illustration of kit pieces being used as parts, Geoff Stippes built these Sn3 (1/64 scale) passenger cars from Model Die Casting HO models and Evergreen styrene sheet and strip (the white parts in the top photo). He used Grandt Line window moldings, commercial end platforms, and duckbill roof sections from the MDC kits to speed up the most ticklish parts of the job. While the results aren't pure scratchbulding, the models are distinctive and attractive.

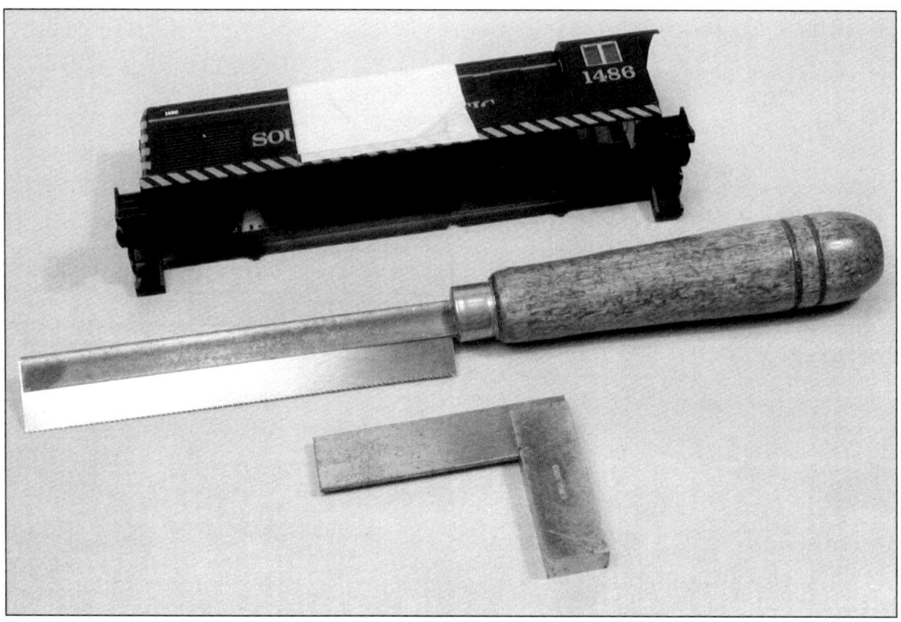

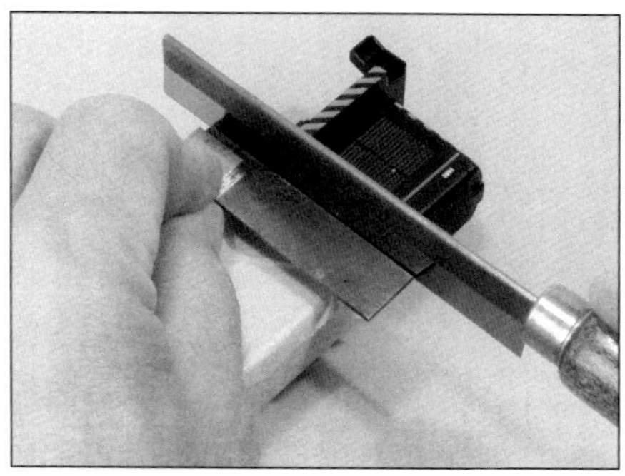

A typical kitbashing project entails making a cut in this HO (1/87 scale) locomotive shell. The job calls for a razor saw, machinist's square, and masking tape. The blade of the saw has been reversed so that it cuts on the pull stroke, offering better control. The tape covers the portion of the shell that will be used in the model, protecting it from scratches.

SIMPLE CONVERSION: STYRENE TORPEDO BULGES FOR U.S.S. *SARATOGA*

by James Kloek

One of the interesting things about ship modeling is that a naval vessel often changed appearance during its service life. To build a properly detailed model of a ship, you need to pick a specific time in the ship's history and build it the way it looked then.

While figuring out how a ship was altered in appearance over time can be fun, the changes present a real challenge for kit manufacturers. The manufacturer must make compromises, and correcting them can be rewarding.

Changes were often made to the hull, where armor belts or torpedo bulges were added or modified. A good example of this was USS *Saratoga*, CV-3, one of the two *Lexington*-class aircraft carriers. Fujimi makes 1/700 scale kits of both: The Lexington is pre-WWII with 8″ gun mounts; the Saratoga is late war, with 5″ gun mounts and a much-augmented anti-aircraft suite.

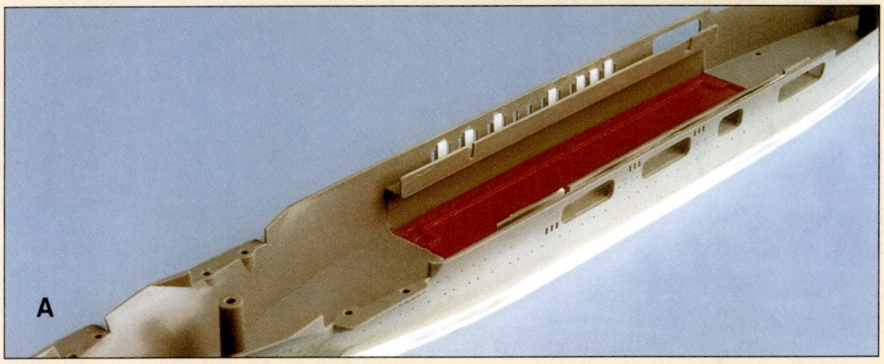

A

Fujimi uses the same hull molding for each kit, the prewar hull. When the Saratoga underwent a major refit early in the war, torpedo bulges were added to each side of the hull. While the one on the port side was relatively small and not readily noticed, the starboard bulge was prominent and obvious. When I started my model of the late war "Sara," I decided to add the torpedo bulges.

The port side bulge was relatively easy, and looked very much like a thick armor belt. Consulting my references, I cut a strip of styrene sheet and glued it to the hull, **A**. A little sanding completed it.

The starboard bulge was a different story. This was wide amidships, and tapered dramatically fore and aft. I started by cutting a thin piece of sheet which stretched the entire length of the bulge.

After gluing the first strip in place, I cut successively shorter pieces of sheet and laminated them on top of each other to build up a sandwich to provide the proper thickness amidships. I then sanded the built up bulge to a smooth taper.

The final step was to fill the seams between the sheets with thinned auto putty, **B** and **C**. Once painted, the effect was quite realistic.

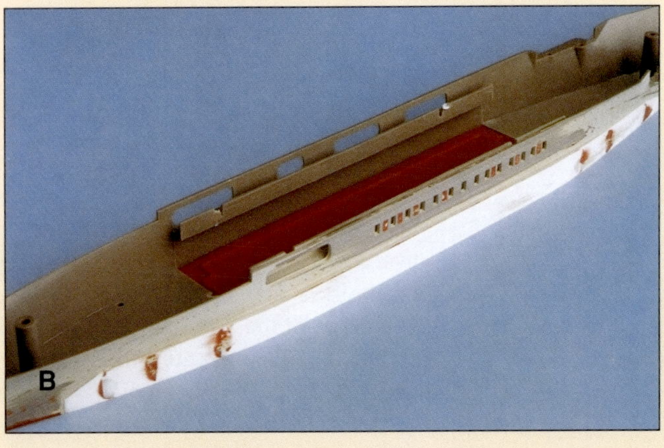

B

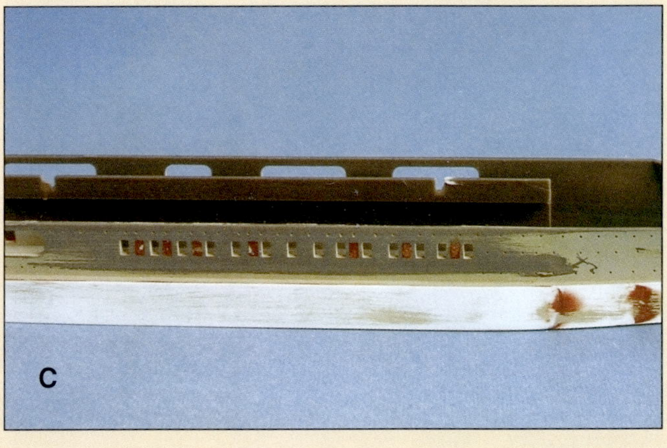

C

CROSS-SCALE KITBASHING

Kitbashing projects can cross scale lines, using parts from one scale to model in another. This happens most often in model railroading and ship modeling, where the real thing can be simply scaled up or down when a different size freight car or boat is needed. Here are three examples, two produced by drawing parts from a smaller scale, and one made by pirating pieces from both a larger and a smaller size.

The inspiration for this distinctive Sn3 (1/64 scale) Alyeska ore car ran on the standard gauge Midland Terminal Railway in Colorado. Bill Adkins cut down the sides of an Athearn HO (1/87 scale) 2-bay hopper and added a top flange of Evergreen strip, Grandt Line Sn3 ladders, and other details.

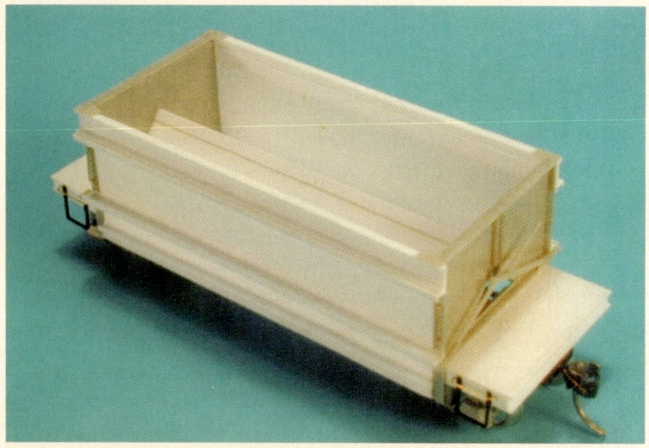

Brian Ellerby used the ends from Grandt Line On2 (1/48 scale) Koppel mine car kits plus Evergreen sheet and structural shapes to build a string of 20 Sn3 (1/64 scale) Yukon Copper ore cars. To convert the Tichy HO scale heavy-duty archbar trucks to Sn3 track gauge, Brian simply shortened the bolsters and glued them to the styrene sideframes.

Modeling an obscure railroad calls for resourcefulness — and kitbashing. Bill Wallace used the body from a Tichy HO ore car kit plus Evergreen strip and Grandt Line details to build this Sn3 (1/64 scale) Lenora, Mt. Sicker 10-ton ore car.

roughed out, check them for overall fit. This is also the place to check that your plan for the conversion is going to work. Then it's time to carefully refine the joints between the major pieces until you can dry fit them together precisely.

Filing and sanding

The best tool for smoothing flat edges is a large file, closely followed by a large wood block covered with sandpaper. A sharp file will make quick work of molded styrene, yielding a smooth, true surface.

Sanding blocks are handy when you have small parts to fit, because the part can be moved across the sandpaper to remove material.

NorthWest Short Line's True Sander is a handy tool for squaring up the edges of kitbashed parts. It's especial-

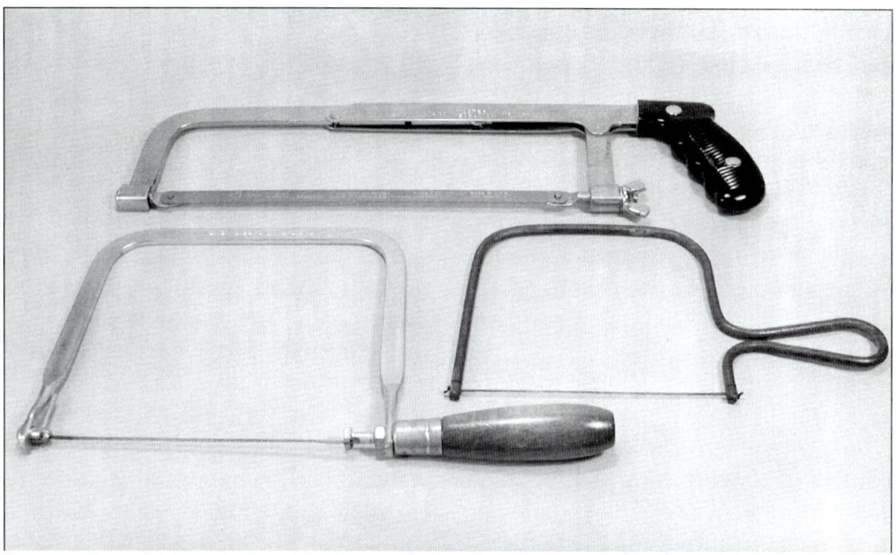

Here's the heavy artillery, saw-wise. The standard 10″ hacksaw (top) is what you'll need to cut through thick plastic sections or laminated blocks. It's also handy if you use PVC pipe sections to build cylindrical shapes. The fine-toothed coping saws are ideal for roughing out openings because the blade can be removed and passed through a hole drilled in the model.

FILES FOR KITBASHING

It's a mistake to think that building small-scale models always calls for using small tools. I count a couple of large, medium-cut flat mill files as essential tools for kitbashing. They're great for smoothing sawn or snapped edges on molded parts, and they remove material quickly — if they're sharp.

An 8″ mill file is long enough to true a cut styrene edge without leaving high and low spots. Its teeth peel away shavings from the plastic surface quickly, like a block plane on wood. The teeth point forward, and cutting action is on the push stroke.

Styrene shavings quickly clog the spaces between the file teeth and stop the cutting action. Clean out the chips with an old toothbrush or brass-bristled suede brush.

Label a couple of big files "For Styrene Only," and keep them where they won't be used for metalwork or home-repair jobs. They will be cut more crisply and stay sharp longer if used only on plastic.

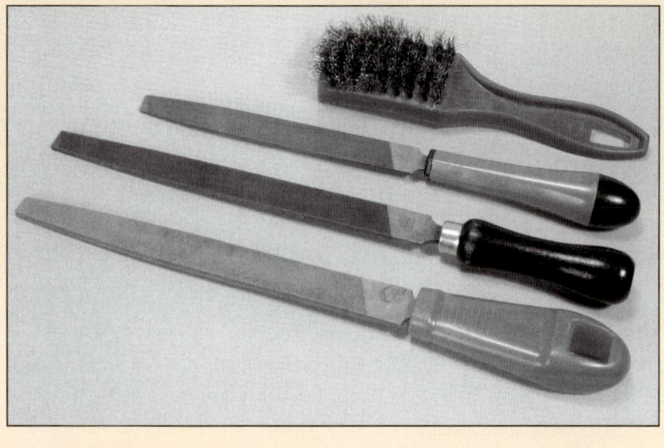

Even if you use them only on styrene, files slowly become dull until they're no longer doing their job efficiently. I make a practice of buying new files every couple of years to compare to the ones that have been on the bench for a long time. Toss out the dull ones! *Bob Hayden*

ly useful for accurately truing the ends of body moldings or other parts with rectangular or cylindrical cross sections.

Adjust the fit of the major parts until the pieces line up properly. As you go, identify areas where reinforcement will be necessary.

Making new openings and plugging old ones

Cut new openings for windows or doors by drilling, sawing, and trimming. If a detailed molded opening can be salvaged from an unneeded part of the donor model, you may want to cut around it, then cut an oversize hole where it's needed in the conversion and fair it into the surface.

Plug unwanted holes with styrene. Small round holes can be filled with a length of sprue or runner whittled to fit, or by enlarging the hole to accept the next larger size of Evergreen styrene rod. When the solvent has dried clip off the excess and sand the plug flush.

Fill larger openings with sheet stock plugs, reinforced from behind. Cut the plug as precisely as possible to minimize the amount of filler needed around the edges.

Removing and salvaging detail

When you hack up plastic kits with a saw, chances are that the surface details may not wind up where they're supposed to be. Remove cast-in detail with a chisel blade (No. 17 or 18) in a modeling knife, or with a chisel made for the job.

NorthWest Short Line's True Sander is especially useful for truing the ends of kitbashed parts. The moveable sanding block on the right side slides back and forth while the model part is held stationary against the top fence and angled guides.

Work slowly, and protect the surrounding surface with masking tape until the last finishing pass. Larger details should first be clipped off with flush-cutting nippers, then the remaining stub removed with chisels and sandpaper.

The photos on page 44 show how to salvage raised surface details from a molded part. First cut around the detail, then sand the backing away until the plastic is translucent. Trim the "flash" with a knife, then sand again to prepare the part for use on the model.

Assembly procedures

Remove alignment pins and lugs from the back of kit parts except in the rare instance where they'll help during assembly. Usually they'll hinder instead of help.

Where possible, overlap parts to help them reinforce each other and align the overall assembly. For example, make the cuts in two halves of an aircraft fuselage at different locations, so the completed fuselage will be strong and straight.

The same goes any rectangular assembliy such as a model railroad buildings. Splice the walls at one location, but offset the joint in the roof to strengthen the model.

Assemble the conversion parts temporarily with masking tape, just as you would do with a kit. This often reveals areas where the converted model will require bracing or stiffenrs for strength.

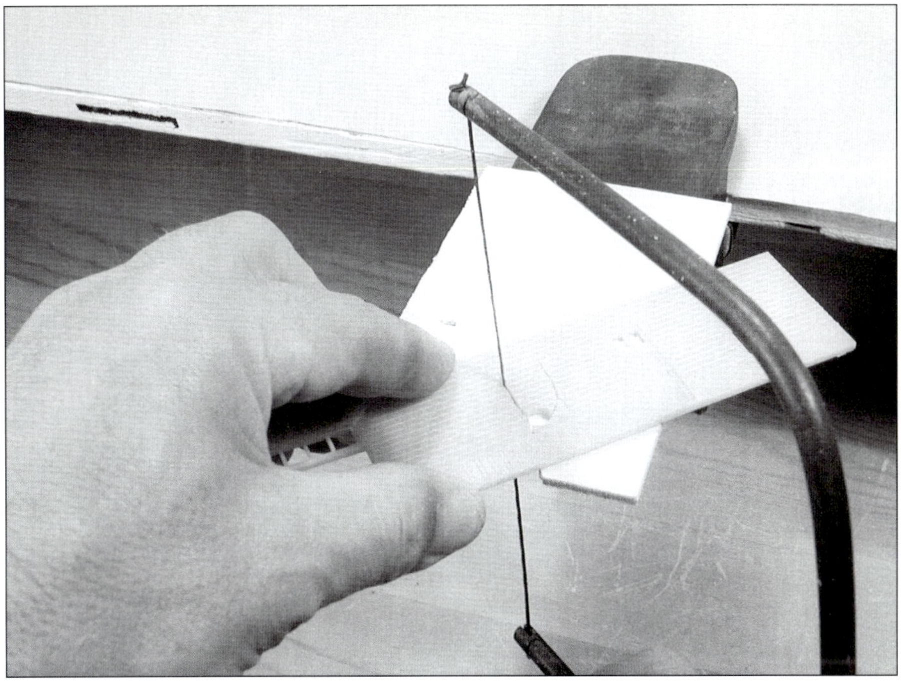

Cutting a new window opening with the coping saw starts with a hole drilled near one corner. After the opening is roughed out, it will be finished with files.

As with any modeling project, think in terms of subassemblies that can be substantially finished, then brought together later in the game to complete the model. Because painting is such an important aspect of every modeling project, my preference is to engineer subassemblies to facilitate painting. Mechanical considerations also apply, particularly in the case of models that must be disassembled for adjustment or maintenance.

Don't hesitate to make up a simple assembly fixture for complicated models where alignment is critical. WWI biplanes, for instance, are best assembled with a jig to hold fuselage

text continues on page 44

FILLING A HOLE WITH SHEET STOCK

A DC-3 I was commissioned to build was one of the odd ones that had her passenger boarding door on the right side, instead of the usual left. The Monogram 1/48th scale kit has the left side passenger door, and is a very good example of a piece meant to be used in the open position only. Closed, the fit was not good.

It was more effective to fill the hole with .020″ sheet stock. The kit door provided a perfect template with which to make my opposite side door panel. I simply glued the new door shape cut from .010″ sheet stock to the proper side and location on the fuselage and allowed the outline of the raised piece to represent the door. *Pat Hawkey*

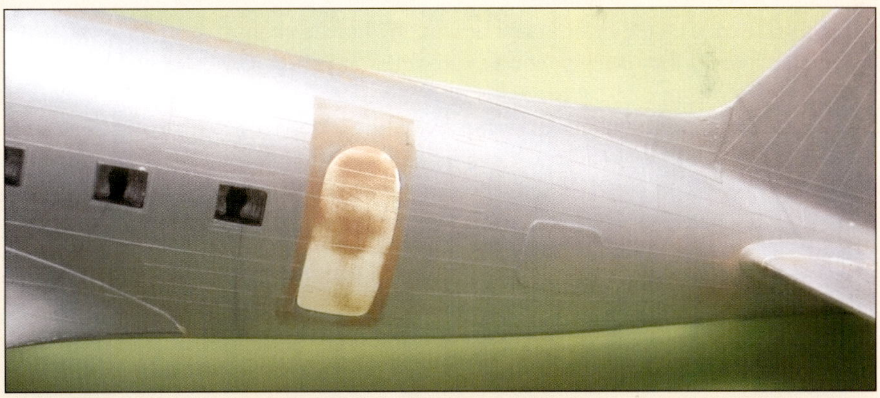

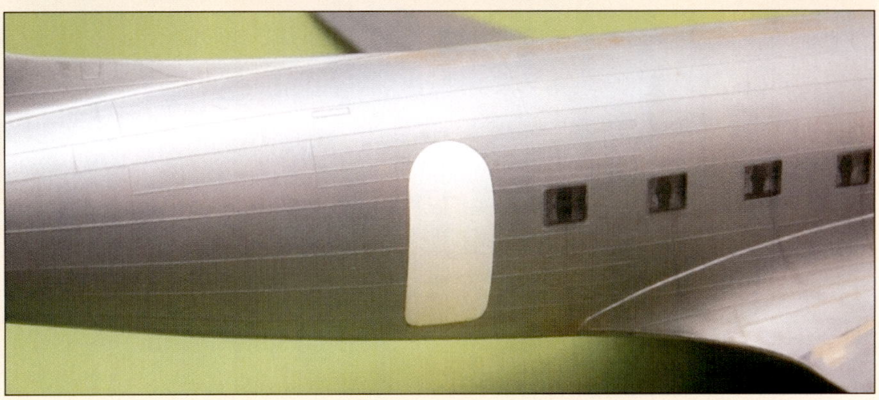

but it was disappointing. The plates that made up the superstructure were at least .060″ thick; in 1/35 scale that's 2-1/8″! The poor little PzI wouldn't even have been able to move that much armored weight on top!

I decided to scratchbuild a replacement superstructure from Evergreen styrene sheet, **A**. Since the real thing was all flat plates, this is not as daunting a task as it seems. And the gun? Verlinden makes a spectacular resin kit of the sIG, and I had one on hand, already built. So, let's cut some plastic!

CASE STUDY IN STYRENE: 1/35 SCALE BISON CONVERSION

by Joe Morgan

One of the first books I remember reading was the huge, red-covered *Picture History of World War II* from Life magazine. An image that really captured my imagination showed a diminutive, boxy German "tank" (at that age I didn't know the difference between a tank and a self-propelled gun) firing point-blank at a building in a French town. I learned later — much later — that the vehicle was a Bison, a sIG 33 150mm infantry gun mounted on a Panzer I chassis.

The Bison conversion was the first attempt by the Germans to provide infantry with close support in an armored vehicle of their own, a concept that culminated in the Stug assault gun. Strictly a stopgap (only thirty-eight were built), the high, boxy superstructure and the heavy gun overloaded the little Pz1, which had only 100 h.p. to lug all that weight around.

I began this conversion by purchasing a resin conversion kit,

Kit construction

The base kit is the Italeri PzI, which has been around forever. Build step 1, the lower hull and running gear, per the instructions. Paint this subassembly dark gray, with the tires "tire black." This is not just flat black — go out and look at the tires on your car. I start with flat black, but mix in dark gray and dark green.

Steps 2 and 3 build the upper hull, and here we begin to diverge from Italeri's plan. First sand the top of the hull, the area with the turret ring, flat. Remove the hinges for the driver's hatch and the turret ring. This is where the floor plate will fit.

Install part B27, the driver's front view slot, but put aside the two front side ones, parts B26. They'll be covered by the new superstructure. Install the rear view slots, but omit the driver's hatches, parts A29, 30, and 31.

Step 2 gives you a choice of two engine-deck variations, one plain, the other with the air vents opened in the hatch at left rear and raised engine vents. Since these vehicles were overloaded, it makes sense they'd need all the cooling they could get: I chose the open version.

Once you've opened the vents, back both this piece and the grille on the right rear with sheet styrene and paint it black to prevent light leaks. (We're leaving off the driver's hatch, so

A

B

there will be light inside the hull.) Finally, draw or scribe a line across the top of the hull, connecting the rearmost angles behind the turret opening, **B**. This will be our reference for placing the new floor plate.

The new superstructure

I used a ruler and drafting square to draw the parts directly on styrene sheet. You might find it easier to draw the parts first on paper (gridded or graph paper is handy for this) and transfer the shapes to the plastic. But you don't have to — I've provided templates, page 43.

I cut the replacement floor plate from .040″ Evergreen sheet. I wanted it to be sturdy, since it is the basis for the whole new superstructure. I used both the score-and-snap method and a razor saw.

It took me three tries to get this right: The resin part, which I tried to use as a pattern, didn't fit the angles at the rear of the hull and around the driver's opening, so I made adjustments and tried again. Something still wasn't right: After checking reference photos I determined that the resin part was also positioned 5mm too far back on the chassis. When correctly positioned, the gun muzzle should overhang the front of the vehicle just a bit. My third try was the charm, with a satisfactory fit.

The correct nonskid pattern for the floor is available as a photoetched brass sheet, Eduard no. 35142, Early German floor plate (dot pattern). I placed the new styrene floor on the Eduard sheet and scribed around it with my knife. I cut the brass with scissors and super glued it to the plastic. It looked great, **C**.

Side plates

The Bison's superstructure sides plates were 10-13mm thick. I used Evergreen .015″ sheet — the conversion table on the back of the package says this equals 1/2″ in 1/35 scale, which is about 13mm, and close enough for me.

The rear third of the sides was hinged. This translates to 20mm on the model. Since I was not going to open these flaps, I cut the sides as one piece, then scribed the joint line on the inside and outside.

I installed four brass hinges from my spares box on the outside. On the inside, I glued a stiffener on the leading edge of the opening consisting of a plastic strip with rivets, also from my spares box. I don't know if this was really there, since I've never found even one clear photo of the inside of the Bison, but it makes sense and stiffens the model part.

Each side has a protruding oval fitting that lines up with the axles of the gun; the axles may have stuck out into them. They're conspicuous on the photos, so I found a couple of oval plastic pieces in my spares box that were roughly the right shape, then sanded and filed them to shape. I placed each on an oval of Evergreen .010″, then added eight Grandt Line rivets — the first of many!

There's another fitting at the upper front of each side plate; it may be a view slot. I made them from .010″ Evergreen, 5mm high by 10mm long, with the corners rounded. Make four and cut out the insides, leaving a frame about 1mm wide. Install one on the inside and one on the outside of each plate, and add six rivets per side, **D**.

Each side plate has a row of Grandt Line no. 153 rivets up the leading edge. Mark the locations with a sharp pencil, 1mm in from the edge and at 1/16″ intervals.

Use a sharp blade to cut the rivet heads off their pins, flush at the back; I cut off about 10 at a time. Put a small puddle of gel-type super glue on a swatch of waxed paper, which seems to retard the setting time of the glue somewhat.

Change to a brand-new blade, and use the point to pick up a rivet, dip it in the glue, and place it on the model. Each side plate has 34 rivets along the front. The rivets are about the size of a flea's head, so take your time or this job will drive you crazy.

Add two Grandt no. 156 rivets near the bottom of each side, then flip the plate over and add a strip of 1.5mm angle down the inside of the leading edge. This stiffens the leading edge and adds some gluing surface for assembly.

The front plates

The lower front plate has a rectangular opening for the dri-

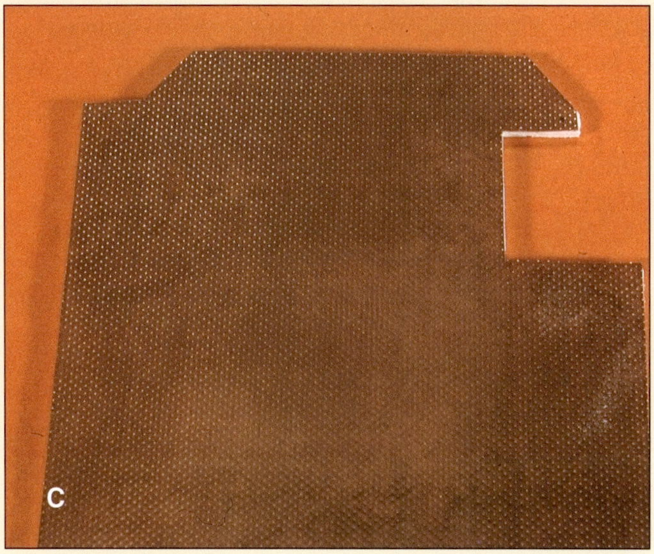

C

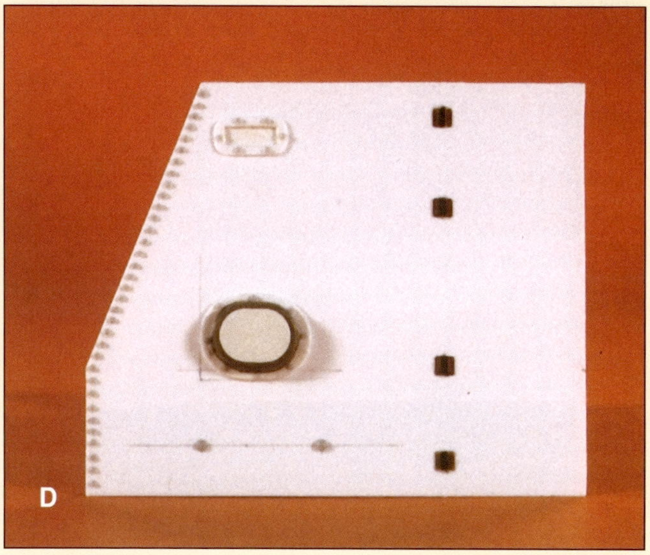

D

ver's vision slot (seeing out of this vehicle obviously left a lot to be desired). Measure and draw the plate on Evergreen .015″ sheet. The critical dimension is the width, which must match the vehicle from fender edge to fender edge. Make it 58mm.

This vertical plate is featureless, except for 54 no. 153 rivets at the outside edges and across the top, plus another 20 in vertical rows beside the opening, **E**.

The upper plate, **F**, has the opening for the gun and a rectangular hinged door that coincides with the hatch in the gun shield for the sight. Make the opening 8mm x10mm, and cut a 10mm x 13mm hatch from .015″ sheet to cover it. The brass hinges came from my spares box.

There are four rows of three no. 156 rivets at the base of this plate. Place the first row in line with the edge of the gun opening, and space the next row 3mm inboard. Some Bisons had a spare wheel mount on this plate; I made this from a scrap of brass runner from a photoetched detail set.

Last — but not least — add 24 no. 153 rivets to each outside edge, then flip the plate over and add either a small angle or I-beam to the lower edge to strengthen it.

Two plates form the inner walls of the driver's opening. Since their thickness won't show on the finished model, I made them from .040″ sheet to add rigidity. Cut, trial fit, and trim as necessary — then add six no. 153 rivets to each leading edge.

Assembling the superstructure

I agonized for a couple of days before devising a way to square up the new base plate and superstructure parts and fit them to the chassis. You can't just plop the floor plate on and go from there: If it's even a tiny bit off, the high side plates will exaggerate and reveal the error.

My solution was to a fixture to align and assemble the parts. It had to be perfectly square, the exact height of the kit superstructure from fender to top, and the exact width of the finished new superstructure minus the combined .030″ width of the two side plates.

Because I didn't want the styrene parts to stick to the fixture when I assembled them, I made it from Plastruct ABS. Ordinarily I don't like to use this material because you have to use Plastruct's special glue to bond it, but in this case that was an advantage.

The fixture, **G,** consists of .095″ ABS sheet sides with 3/8″ ABS I-beam crosspieces. The sides are 12mm high x 65mm long, and the width of the completed assembly is 57mm. One of the I-beams is set back 12mm from the front to accommodate the inner front panels.

I began assembly by gluing the inner front plates to the vertical front plate. Then I glued one side plate to the floor plate, sitting on top of the fixture. Next I added the vertical front plate with its two inner plates, followed by the second side plate.

The last piece to be added to the assembly was the upper front plate, **H**. Then I took a deep breath, removed the superstructure from the fixture, and tried it on the chassis. It fit! Photos showed small pieces of angle iron where the front of the superstructure touches the fenders. I used the smallest Plastruct angle for this, cut to the fender width, **I**.

The gun

My Verlinden gun kit was already assembled, so I had to modify it much like the real thing to adapt it to the chassis. I removed the wheels and brake drums, then made new stub axles from a metal bolt and attached them to blocks of 1/4″ x 1/4″ balsa with brass strips and Grandt Line nut-bolt-washer castings, **J**.

The rear of the gun trail had to be sanded a bit to achieve a good fit on the rear hull. Photos showed some sort of fittings holding the gun at the rear, but not very clearly, so I made up the fittings from stuff in my spares box.

The gun shield needs to closely follow the angle of the front superstructure plate; as built, it is angled much more steeply. I cut about 1/4″ off the lower shield support. The upper supports will flex enough to allow you to push the gun into place as far forward as it will go. This will give the shield the right angle, or pretty close to it.

Final assembly

The solid plastic Italeri headlight was not very inspiring, so I drilled it out to accept an MV Products no. L128 lens. I replaced the exhaust shroud (part C47) with a box made from .010″ Evergreen sheet and brass screen, then formed it around a rod the same size as the muffler. This was a big improvement.

Part A37, the siren, was replaced with one from a Tamiya

E

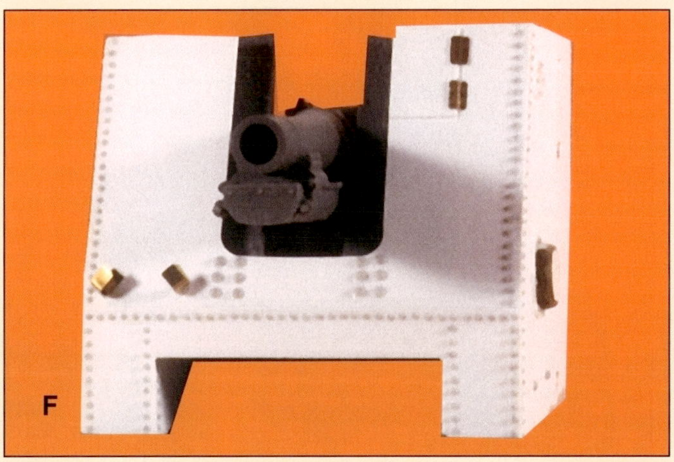

F

PzIV accessory set, and I used all the tools except the engine crank and jack from the same set. I shortened the air intake, part A65, to fit, as the real ones had to be shortened. The model was painted with my own mix of Panzer Gray acrylics, and dry-brushed with successive lighter shades of gray.

This was a fun conversion. Working with Evergreen sheet was a snap because of its texture — soft and easy to cut — and compatibility with conventional plastic cements. The large range of thicknesses made it possible to build the model with scale size plates, a definite plus. With a little help from Italeri and Verlinden, you, too, can have a pet Bison!

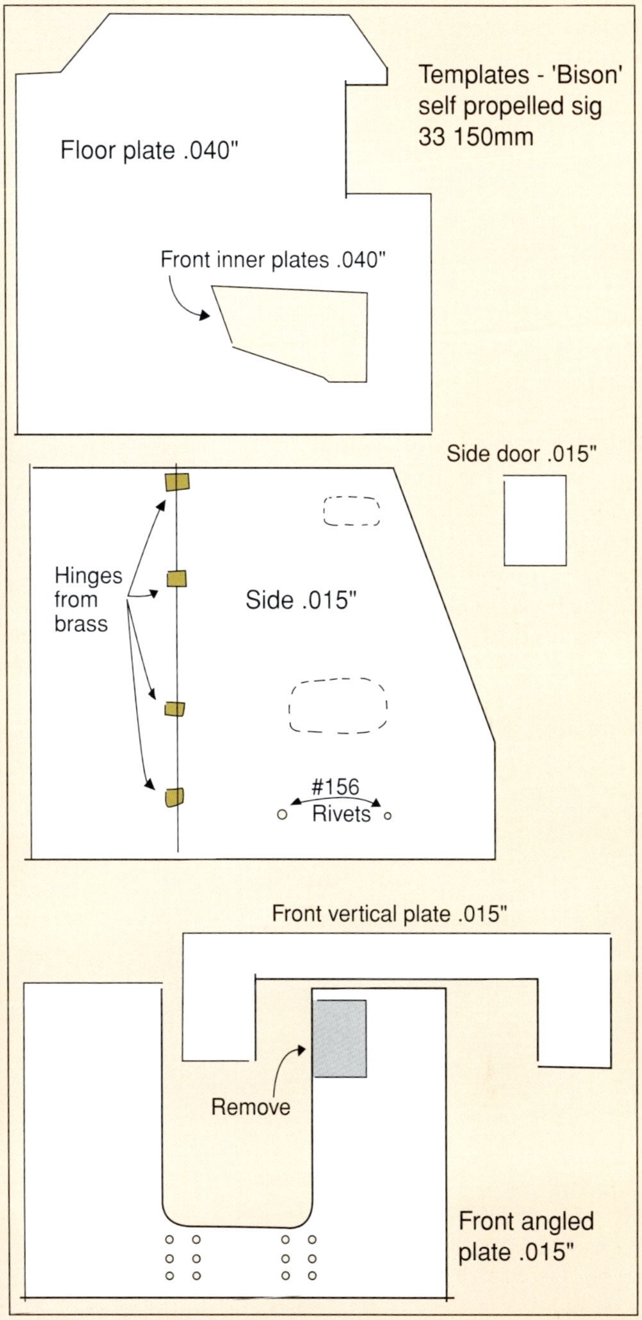

G

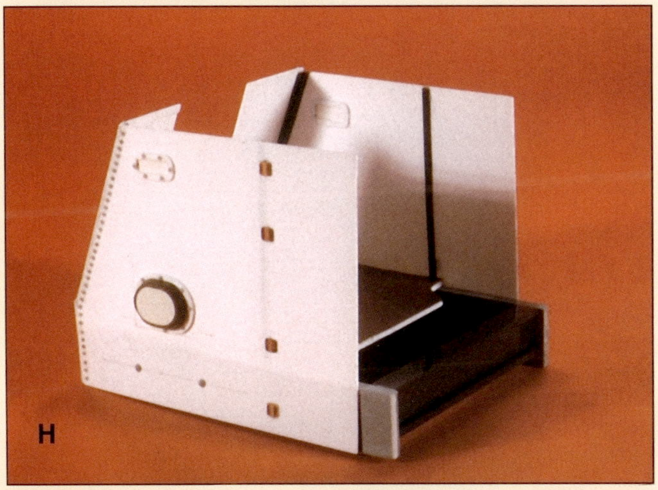
H

I

J

and wings in position while the glue sets and the struts and rigging are installed.

Strength is an important consideration in kitbashed models, since the inherent rigidity of the original kit parts is compromised when you saw them apart. In addition to reinforcing strips and gussets on the back of the parts, think about inserting wire pins or brass or plastic tubing to ensure alignment, add strength at joints, and key parts together.

Removing molded-in surface detail (in this case, so it can be replaced by free-standing wire parts) calls for care and patience. Micro-Mark offers this detail chisel specifically for the task; the No. 17 hobby knife blade also works well. It's best to slice away such details in several thin cuts to prevent gouging the underlying surface.

Murphy's Law of Kitbashing

We started this discussion of kitbashing with the thought that it's a technique for producing models that are not available as kits. Be aware that one of the many Murphy's Laws of modelbuilding is: About the time you paint an extensive conversion and begin to apply decals, a kit will be announced!

SALVAGING SURFACE DETAILS

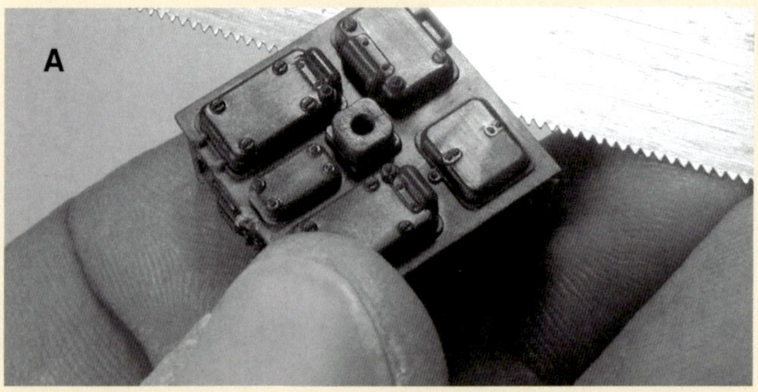

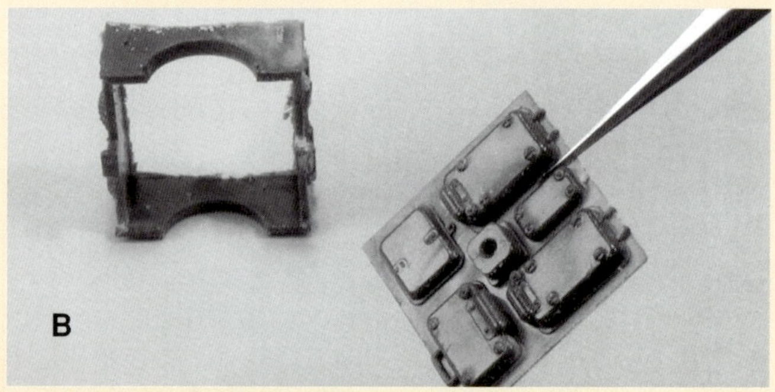

You'll often want to salvage surface detail from injection-molded parts for use on a kitbashed model. Start by whittling down the area around the details to a small area, then saw through the plastic behind the details, leaving a wafer of plastic with the parts you want on the surface, A.

Next, rub the wafer on the surface of a sanding block to thin the backing. As the plastic wears away, switch to wet-or-dry sandpaper, used wet, to sand it paper-thin, B.

Cut through the thinned backing with the point of a new No. 11 knife blade, then return to the wet sanding block to remove the final thickness of the backing from the part, C. Now you can apply the part to your model just as if it had come in a little plastic bag!

This beautiful 1/48 scale Spitfire Mk.IX is the work of Bob Steinbrunn. He started with an Arii Mk.VIII kit, then added dozens of photoetched cockpit parts and details made from sheet and strip styrene.

Chapter 7

Detailing Techniques

HOW MUCH DETAIL IS ENOUGH? That's a question that every modeler has to answer on every modeling project, and it's never an easy one. Details range from panel lines scribed into the surface of the model, to raised bits like rivet or bolt heads, to extensive interior treatments.

With kit models, much detailing involves first removing inaccurate or oversimplified details, then replacing them with after-market or scratch-built items. The hardest aspect of this work is often getting rid of the unwanted parts without damaging the underlying surface of the model.

If you are new to modeling, the best

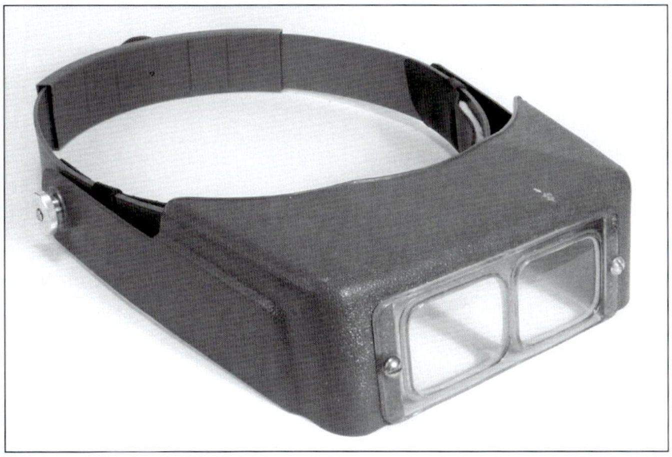

Modelers of all ages have been wearing Donegan OptiVisors for decades to help with close-up work. A pair of inexpensive magnifying glasses from the drugstore will also do the job, but they won't flip up out of the way like the visor.

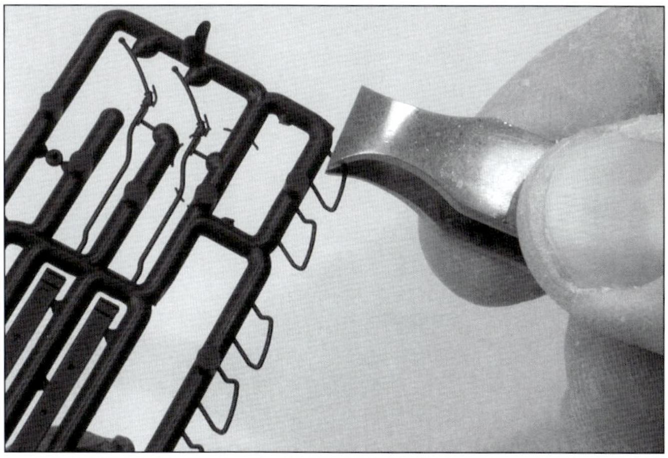

De-spruing tweezers have sharpened jaws to cut through the gate that connects the part without twisting or bending it. You'll want a pair to deal with very delicate molded styrene detail parts such as these tank car handrails.

THE FINE ART OF STRETCHING PLASTIC

by Pat Hawkey

Gun barrels and pitot tubes are obvious candidates for replacement on most kits. Whether they're molded separately or attached to larger portions of the model, you've got two mold lines running the length of the object to clean up. If they are molded on the airplane, **A**, you've got to somehow avoid breaking them off during construction.

I find it far easier to clean these things off (along with most antennas) and use them as patterns for replacement pieces to be added at the finishing stage.

With guns and pitot removed (but not thrown away), and leading edge seam sanded smooth, holes are drilled to accept replacement guns and pitot tube, **B**. On this Monogram 1/48 scale P-47, this also allowed me to locate the guns accurately. Incorrectly molded on the centerline of the wing leading edge, the Thunderbolt's guns were in fact staggered across the dihedral of the wing so they'd be parallel to the ground when the plane was parked.

While hypodermic needles make great gun barrels and pitot tubes, you can't always find the exact diameter you need, and stainless steel is not easy to work with. I make my own plastic hypo tubing by stretching Evergreen tube the same way I stretch sprue, **C**. Rotate a length an inch or so over a candle flame until it becomes shiny and softens.

Pull on the opposite ends, **D**, and you get plastic thread with a hollow center. The speed with which you pull and the length of area you heat will determine how thick or thin a segment you end up with. Keep the ends taut until the tube cools to ensure straightness. A little practice and you'll be able to achieve any thickness you want.

The new items test-fitted in place, **E**. The kit parts were used only to determine the correct length of the replacements.

Gun tubes are simply lengths of Evergreen rod stock cut carefully to get a perfectly flat end. (When finished, a dab of flat black paint on the ends will adequately represent a hollow tube. Fussier modelers will want to go with stretched tube.)

Pitot tube is stretched styrene tube with a length of stretched kit sprue threaded through the center to get the stepped front portion, and the blade on top is a bit of .005″ sheet trimmed to shape. Once fitted, the delicate parts are removed to be installed permanently at the finishing stage.

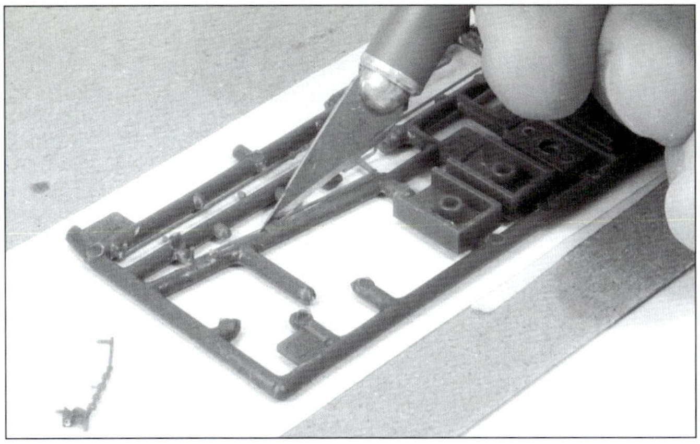

Fold a piece of masking tape with the sticky side up to grab small details as you separate them from the casting sprues. The adhesive surface will keep the parts from flying merrily across your workbench.

The interior detailing on this 1/48 scale Spitfire Mk.IX had to be installed, painted, and virtually 100 percent complete before the fuselage could be closed up. Model and photo by Bob Steinbrunn.

way to learn detailing techniques is to follow a project in a magazine, or one of the sidebar projects in this chapter. One way to get started slowly is to add at least one or two extra details to every kit that you build, to make it distinctive and your own.

Resist the temptation to add detail where it won't be seen, or where special effort will be required to notice it. Instead, devote extra time to detailing the parts of the model that viewers — and you — will see and appreciate.

Good reference material is essential for accurate detailing. Even freelanced models depend on references for convincing detail, and historically accurate models demand photos and drawings.

Yet another Murphy's Law of modeling is that photos of the "missing" side of the real thing will always be

text continues on page 51

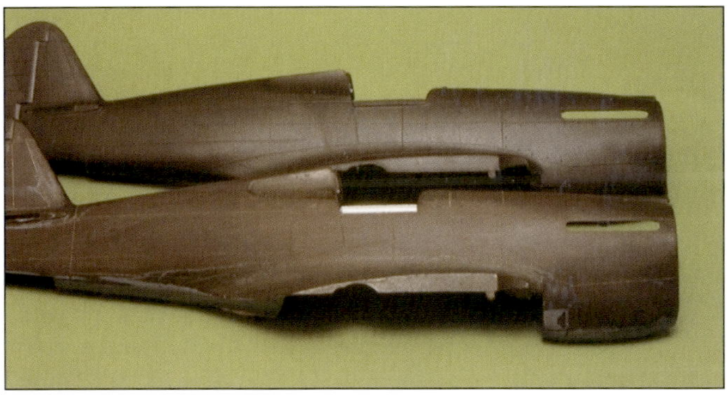

Monogram's vintage 1/48 scale P-40B/C kit gave you the option of open or closed canopy, but the bottom frame of the sliding portion was molded in the closed position on the fuselage. Posing the canopy slid back would therefore give you the unlikely situation of the bottom edge closed, and the clear bit open. The bottom fuselage half shows the problem area removed and replaced with channel stock to represent the railing that the canopy frame rollers roll along. — *Pat Hawkey*

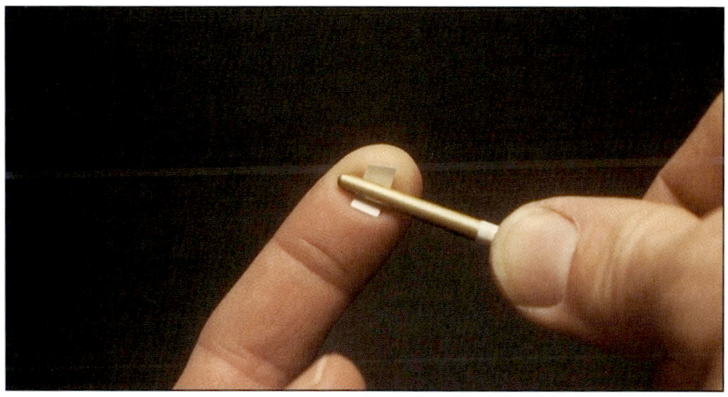

(Above and below) You don't have to heat thin styrene to get it to form simple curves. I needed to make open cowling gills for an Accurate Miniatures 1/48 scale TBF. I cut the gill shapes from .010″ sheet stock, then rolled the handle of a paint brush over them against my fingertip. This curved them evenly to match the cowling. — *Pat Hawkey*

The strength and thinness of photoetched metal parts yields astounding scale appearance. Here's the 1/48 scale Spitfire cockpit in progress. Photo and model by Bob Steinbrunn.

The pewter rudder had a rough texture and lacked detail. I replaced it with one made of three laminations of .040″ styrene sheet. Styrene is easy to work with and can support the carving and filing of complex shapes such as the sections of the three-fingered hinge seen here. The new rudder has two brass rods to mount it to the hull; the aft rod serves as a working hinge pin.

CASE STUDY IN STYRENE: UPGRADING USS KIDD IN 1/192 SCALE

by Robert N. Steinbrunn

My model of a World War II Fletcher-class destroyer, U.S.S. *Kidd* (DD661), shows what can be done to improve, enhance, and detail a warship kit with styrene. This famous destroyer had an illustrious history in the Pacific and now resides in Baton Rouge, Louisiana, as a memorial open to the public.

I started with a 1/192 scale kit by Bluejacket Ship Crafters. The kit provides a carved basswood hull, many smaller wood parts, Britannia pewter cast metal parts, and several frets of photoetched brass.

I added thousands of extra parts made from styrene — not only to detail the model, but also to make it look more like the steel warship it represents.

The ship is being assembled by the "building block" method, where each subassembly is built and finished prior to attaching it to the model. Evergreen styrene takes all hobby paint finishes well, including lacquer-based ones, provided they're lightly airbrushed on.

Although I've been working on this model for four years and about 2000 hours, I'm still not finished with it. When it's finally done, I believe it'll be something I'll be proud of.

Using plans, dividers, a sharp drafting pencil, and Evergreen .005″ sheet styrene, I duplicated the in and out strakes of steel plating on the real hull. The plates were attached to the hull with 3M spray adhesive. This thin sheet is able to take compound curves with just a little burnishing effort.

A close-up view of the inverted hull, showing where styrene sheet and strip were used to make plating strakes, joint doublers, propeller guards, and other details.

Styrene lends itself well to multimedia construction. Here N gauge model railroad photoetched stainless steel walkway grating is being attached. I'm applying super glue with an tool made of a pencil with a thin length of steel wire inserted into the eraser end for extra-fine control.

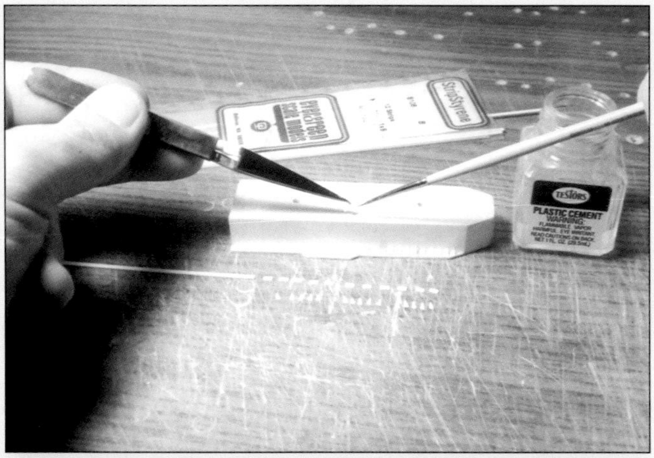

I plated over the main deck and forward, amidships, and aft deck houses with .005″ sheet styrene. This makes them look more like steel and eliminates having to use wood fillers or varnish to hide the wood grain. Super glue attached the styrene to the wood, each surface was covered, then trimmed close when the glue had dried. On a typical square shape I would plate over the north and south faces, allow the glue to dry, trim, then plate the east and west faces. This prevents you from having to glue one surface adjacent to another where the glue is not yet set. Here I'm adding triangular gussets to the aft deck house deck overhang. The .010″ strip is first cut into proper-length rectangles, then each rectangle is cut diagonally into two halves. I bonded the gussets in place with liquid cement applied with a #00000 paint brush.

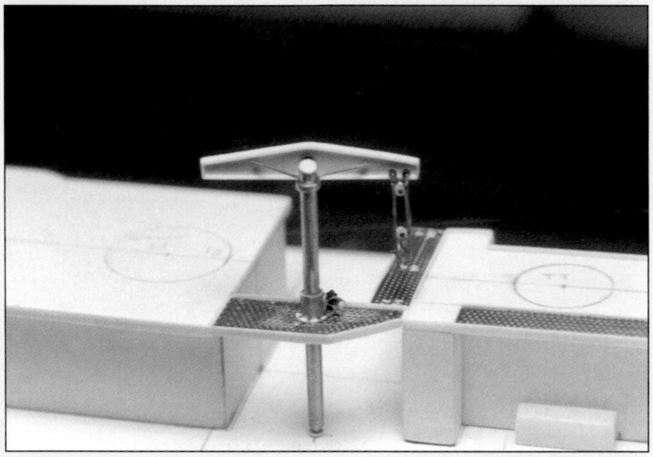

This scratchbuilt torpedo crane was made of styrene strip, brass tubing and rod, and stainless walkway grating. Pencil marks show where fittings and structures will be located.

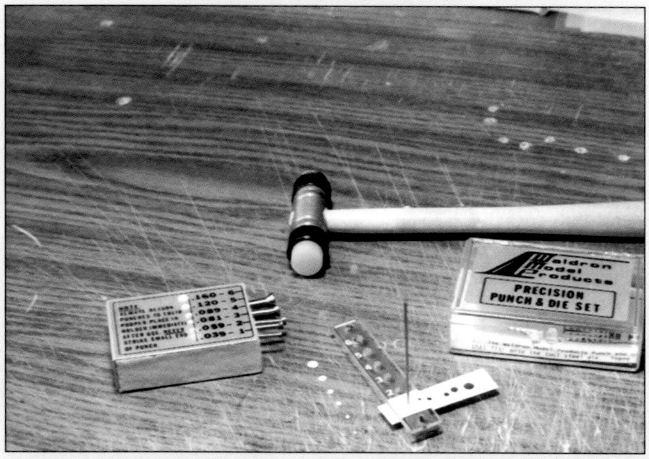

This Waldron precision punch and die set is being used to punch out crisp styrene disks of various diameters and thicknesses for details and fittings.

Strip and sheet styrene was used to detail the 5″/38 main gun mounts, including the mount captains' blast hoods.

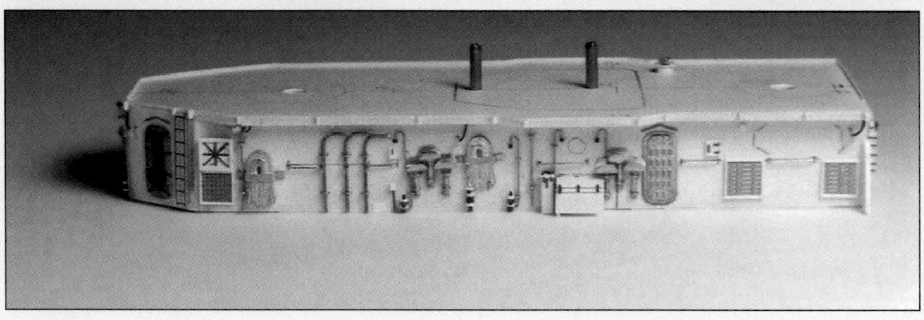

The aft deck house, covered in .005″ styrene sheet and extensively detailed. Note the use of sheet and strip styrene to create ammunition and fuse lockers, vent housings, waterway bars, and stanchion sockets.

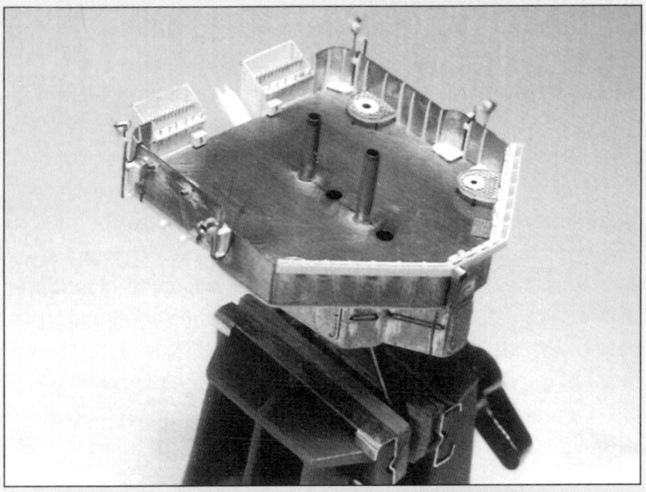

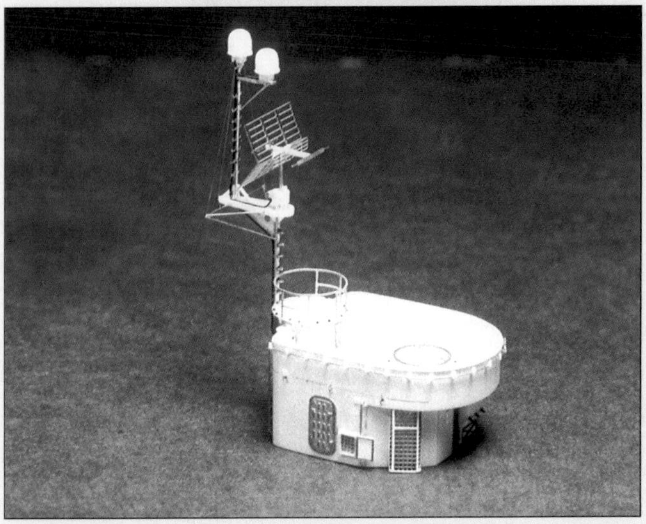

The aft 20mm gun house has been covered in .005″ sheet and various details added: vent housings, ammunition shell case cage, tarpaulin clips, and antenna mast deck. I chucked square stock in a hobby lathe to create the mushroom-shaped DBM electronic countermeasures antennas.

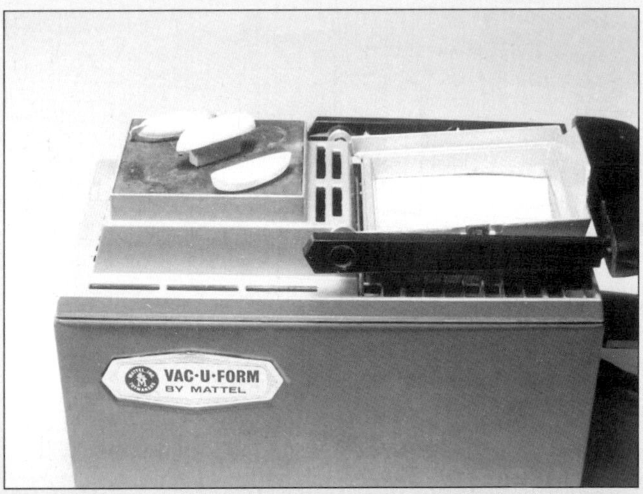

A Mattel Vac-U-Form machine was used to create new 26-foot motor whaleboats to replace the cast metal ones in the kit. I vacuum-formed .010″ sheet over the metal parts.

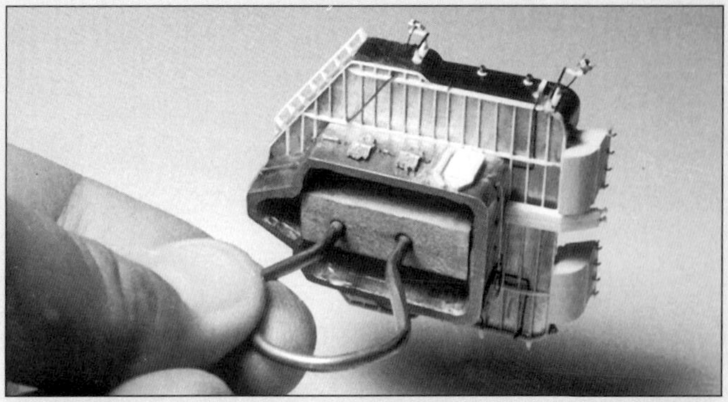

(Left and above) Top and bottom views of the revised bridge deck, showing the use of sheet and strip styrene to create the wind-deflecting venturi on the face of the bridge, the signal flag holders, and various other details.

The kit's cast-metal funnels were shaped, hollowed, and have sheet styrene ductwork inside.

The whaleboat decks were made of .010″ sheet. Using a straightedge and a scribing tool, I was able to simulate the planking used in the real deck's construction.

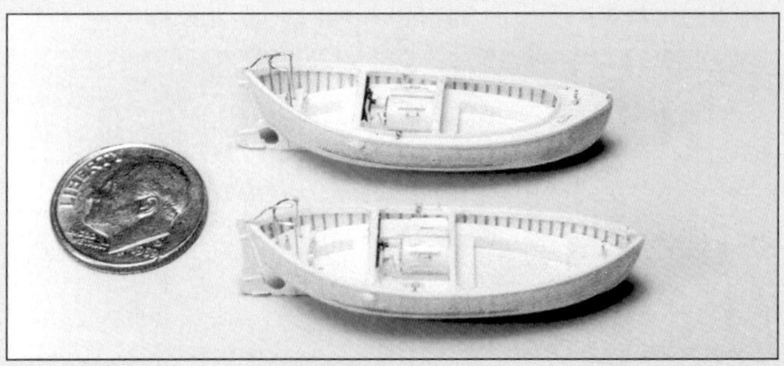

The completed styrene whaleboats; each contains 182 parts.

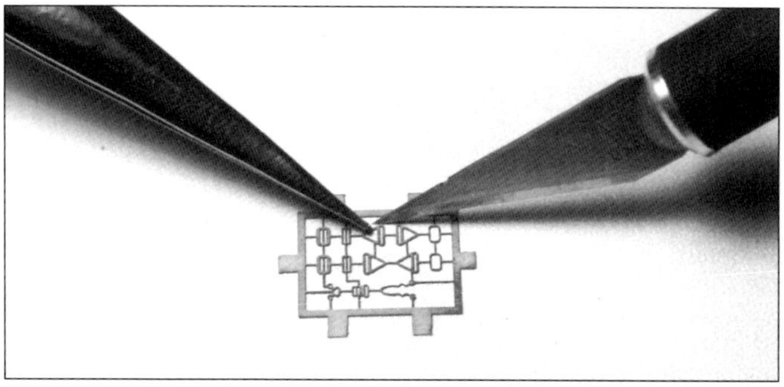

Separating photoetched parts from their fret calls for a sharp knife and patience. Safety glasses are a good idea, too, since the pieces can go flying. (Left) Parts supplied on a flexible backing should be dipped in lacquer thinner to free them.

published in a magazine the month after you complete the model!

I make a habit of keeping my best two or three references on or near workbench during construction. Even on a straight-from-the-box kit model, I jot down a list of possible details as they occur to me. I don't always add them all, but the list means I won't forget something important.

Detailing with styrene parts

Never underestimate how much interest a few simple boxes or panels made from strip stock can add to an otherwise plain model. Nothing could be simpler than detailing a styrene model with styrene parts, whether you make them yourself or buy them in clear plastic envelopes. Application to the model is easy, and the key is to use the absolute minimum of solvent required.

The trend in commercial detail parts is for them to be more and more finely molded and delicate. This means they can be tricky to remove from the molding sprue, and tricky to handle. (It also makes them tricky to see. Regardless of your age, you'll occasionally want some optical help. Binocular magnifiers such as the OptiVisor provide hands-free assistance, and at least a little depth perception. Good lighting helps, too.)

A pair of de-spruing tweezers will help you separate small parts cleanly from their molding frames. Placing the sprue on top of a piece of masking tape, sticky side up, will keep parts from flying off into never-never land as you detach them with a sharp knife blade, as will slipping the sprue into a clear plastic sandwich bag and cutting through both bag and part.

An important consideration in adding detail is when to do it. Sometimes the sequence is obvious — interior framing in an aircraft fuselage has to be installed and painted before the fuselage is closed up — but more often you'll need to think it through. Complex projects call for assembly of detailed subassemblies, then uniting them.

Details that will be easy to damage or that will make the model hard to handle should be installed late in construction, perhaps as painted subassemblies after the main portion of the model has been completed. Antennas of all kinds fit into this category.

White metal and resin details

The main concern in working with white metal and resin detail parts is removing molding flash and cleaning up the castings for use. Thin flash and molding lines can be trimmed away with a sharp knife, but thicker sections call for files.

Cast metal parts may lack the crispness of injection-molded plastic, and the sidebar on ship detailing in this chapter shows how Bob Steinbrunn extensively reworked white metal parts for his model of USS *Kidd*.

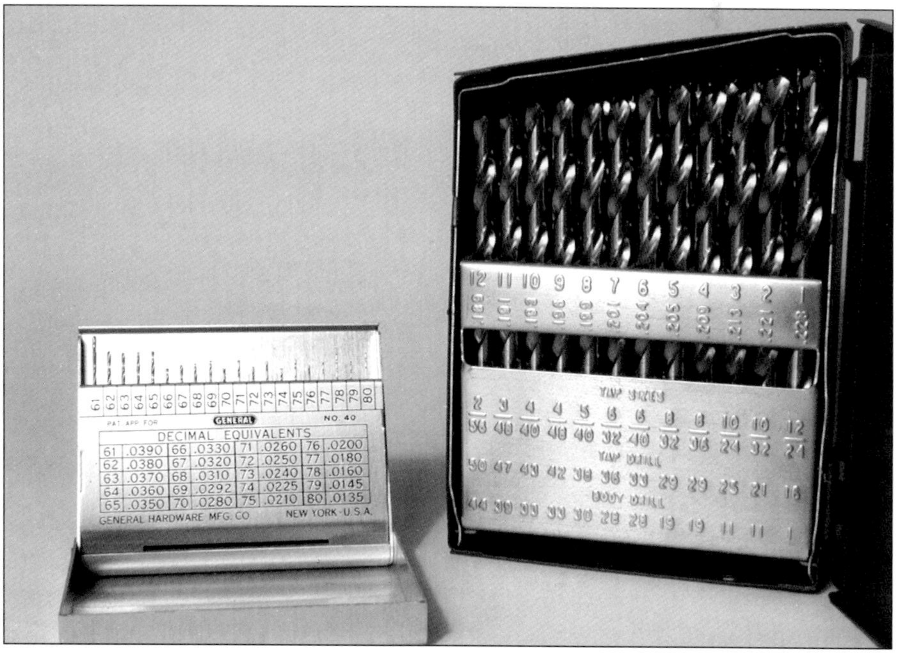

The set of small drills on the left, No. 61-80, will handle most of the holes you need for detail work. The No. 1-60 set on the right is a luxury, but well worth the cost if your modeling calls for holes in that range of diameters.

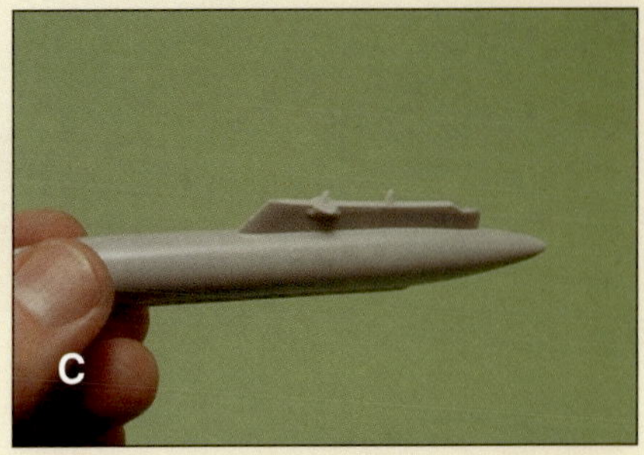

A BETTER WAY TO MOUNT UNDERWING STORES

by Pat Hawkey

This Hasegawa 1/48 scale F-4 Phantom, **A**, has a problem typical of many modern jet kits. Underwing weapons racks and drop tanks come with molded-on pins to fit into corresponding holes. These pins are usually nothing more than pimples that will never actually hold the parts in place, and a mold line invariably runs through them.

You need to clean up the mold line on the inboard surface to get the rack to fit snugly to the bottom of the wing, and the pimples only get in the way. My solution is to replace them with Evergreen styrene rod, which can then be inserted into the locating holes deep enough to get a firm fit, with or without glue.

Rather than finding rod to match the existing holes (and because the wing is a little thin in this area anyway), I filled the existing holes with small diameter tube, securely anchored on the inside surface of the wing with a dab of thick super glue, **B**. When this dried, I shaved off the excess tube on the outer surface with a single-edge razor blade.

The molded-on bumps were shaved off, their locations marked by holes pressed in with a needle point, then holes were drilled to take the styrene rod. Not only are you providing yourself with a strong positive fit with this modification, you'll also have a handle when it comes time to paint the part.

The same diameter drill bit used to make the holes for the pins is used to ream out the new locating holes to ensure a perfect fit, **D**. Look Ma, no hands! No glue either, for that matter, **E**. The drop tank and rack are now held securely in place by the new pins alone.

Metal and resin parts require super glue or epoxy to fasten them to a styrene model. Use epoxy for large, heavy parts, and consider reinforcing joints with wire pins where a heavy white metal part must be attached.

Photoetched parts

Detail parts made by photo etching thin metal sheet (the technology is similar to making printed-circuit boards) offer crispness and realistically thin cross sections. They are especially useful where scale-thickness plastic or resin parts would not be strong enough to withstand assembly or handling.

Most photoetched parts come attached to a frame, or fret, and must be separated from it. To avoid bending the parts, hold them against a metal or glass surface and cut them with a sharp blade. (This dulls the blade, so you'll need to replace it often.) Work on a sticky surface or hold the parts with tweezers to keep them from flying off.

A few photoetched sets come attached to a flexible plastic backing sheet. This means you don't have to cut them free, but you do have to soak the part and its backing in lacquer thinner to remove the backing and adhesive residue.

The etching process may leave a slight burr on the edges of the parts. True the edge (especially if it will be a gluing surface) by light filing or sanding. Sanding sticks are handy for this, and for removing the sprue-like nubs where the parts were

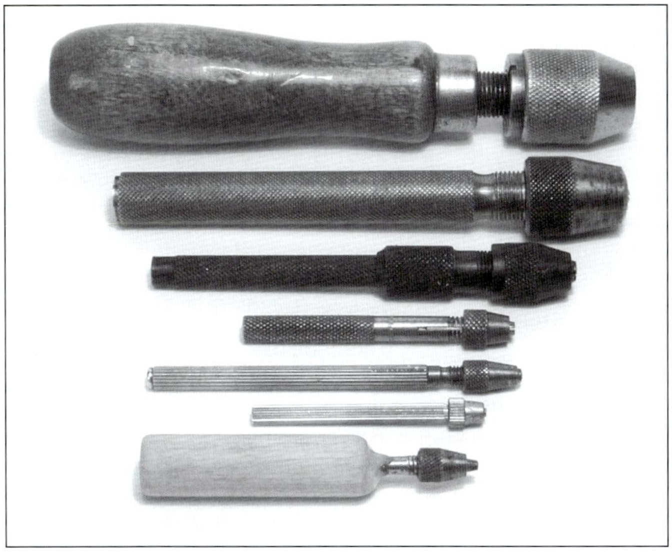

Here's the pin vise collection of a modeler who never saw one he didn't like. The three large models at the top are suitable for bigger drills; the ones below them are the right size for wire-size drill bits. The homemade job at the bottom consists of a micro drill bit adapter chuck for a motor tool epoxied into a piece of 1/2″ dowel. Weighing only two tenths of an ounce, it's perfect for drills in the No. 75-80 range.

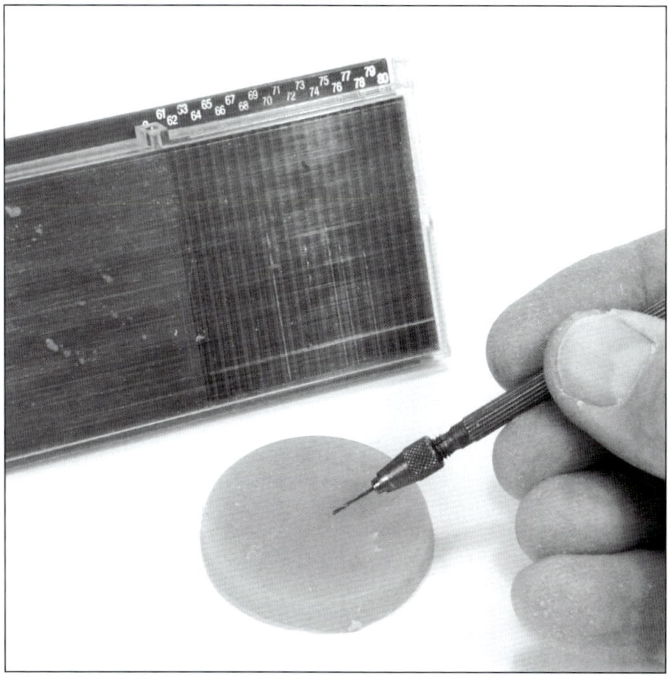

A cake of beeswax from the sewing store makes an excellent lubricant for small drill bits. Just drill the trip of the bit into the wax, then go to work on your model.

INTERIOR DETAILING WITH STRIP STOCK

by Pat Hawkey

You don't have to look far into the past to find aircraft kits (especially 1/72nd scale) that lack the current state-of-the-art detailed interiors — if they have any interior detail at all. The subject pictured here is a fine example. Tamiya's 1/48th scale Lancaster, originally issued in the early 1970s, is not likely to be replaced by a better rendition any time soon. While overall it's a superb kit, the interior is decidedly sparse by today's standards.

The model I was asked to build required that the rear crew-entry hatch be positioned open. If you're going to open a door like that, there has to be something to look at when you peer in!

Having no photos of a Lancaster interior, I simply used the lines of rivets on the fuselage exterior to give me locations for interior ribs and stringers. I determined how much area would actually be seen on the finished model (no sense building up the whole thing when only 20 percent could ever be visible) and cemented vertical square rib strips to establish the borders of my detailing project.

Then, using the model's external rivet lines as spacing guides, I repeated the procedure until the space between the first two strips was filled in. The horizontal stringers were more involved: Each segment between a pair of ribs had to be measured, cut, and individually inserted, **A**. The rivet lines molded on the outside of the fuselage gave me the number and location of stringers required.

Along with the skeleton framework, I included a floor of .030″ sheet, an ammunition trough made from thick rectangular strip, a horizontal stabilizer spar from a box shape found in my spare parts bin (throw nothing away!) and odd bits of equipment just to busy things up a bit, **B**. I don't claim 100 percent in the accuracy department, but it looks good, especially when viewed from outside, **C**.

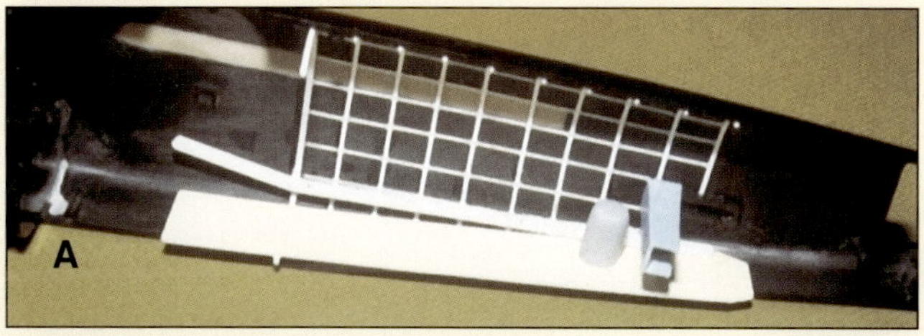
A

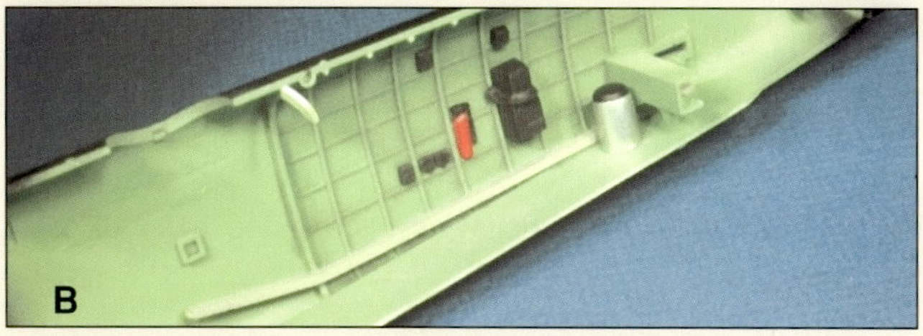
B

C

attached to the fret.

Bending photoetched parts can be tricky. The thin metal fatigues easily: Bend it more than once, and it will break.

Even when they are intended for a specific kit, photoetched parts require repeated test fitting. Most often you'll have to trim away areas of the kit to allow the parts to fit properly.

Super glue is the adhesive of choice for attaching photoetched parts, and you'll often need an accelerator to make the glue kick while you hold the parts in place on the model.

Drilling holes for details

Securely installing details like antennas, railings, and wire hand grabs calls for drilling holes. The combination of very small drills and styrene means you'll be drilling by hand, both to minimize drill breakage and to keep the plastic from melting.

The tiniest twist drills come in sets numbered from 60 to 80, with the highest numbers being the smallest sizes. You'll want a complete set, with an index to store them, and you'll probably want a few spares on hand of the sizes that you use — and break — most often. Small drills are virtually impossible to sharpen; when they dull, discard them.

In addition to drill bits, you'll need a couple of pin vises to hold them. These are small collet chucks with handles. Some come with interchangeable collets to hold different size drills.

Drills smaller than No. 60 are deli-

A

B

BETTER DETAILS FOR AN ARMORED CAR

by Joe Morgan

The 1/35 scale Commanders resin kit of the World War II Hungarian armored car, the Csaba, is the only kit of the vehicle. I upgraded it with styrene details.

The side view, **A**, shows the scratchbuilt Evergreen sheet toolbox on the rear fender and the antenna mounts which replaced improperly shaped parts on the kit. The mounts were round, where my references showed them to be square.

The resin ones were removed and new ones, **B**, fabricated from Evergreen with Grandt Line bolts. The antenna supports were made from modified Tamiya SdKfz 232 parts; the antenna itself is brass wire.

The kit turret hatch didn't fit the turret opening, so I made a new one from styrene sheet, **C**. The brass hinges, **D**, came from my spares box. The Hungarian markings on the painted model were made from 1/72 scale aircraft decals

C

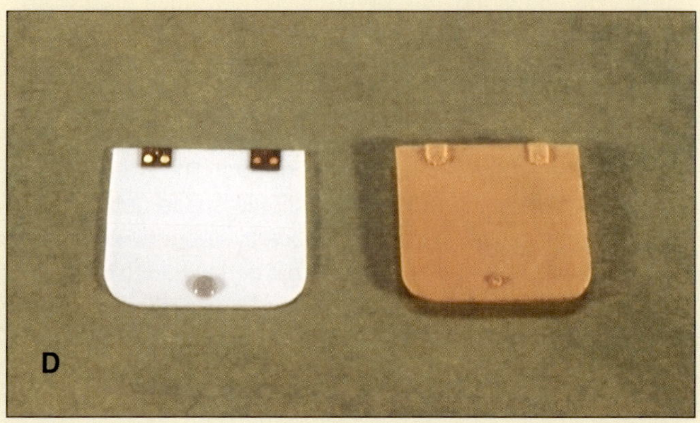

D

The tiny details on the engine of this 1/48 scale He 162 (a WWII German jet) look complex, but they were added one at a time. Wire strands, fine solder, and styrene strip and rod make up most of the added parts. Bob Steinbrunn had to paint the pieces as they were attached, because each successive layer made previous parts inaccessible. Note how the contrasting colors make individual detail parts stand out.

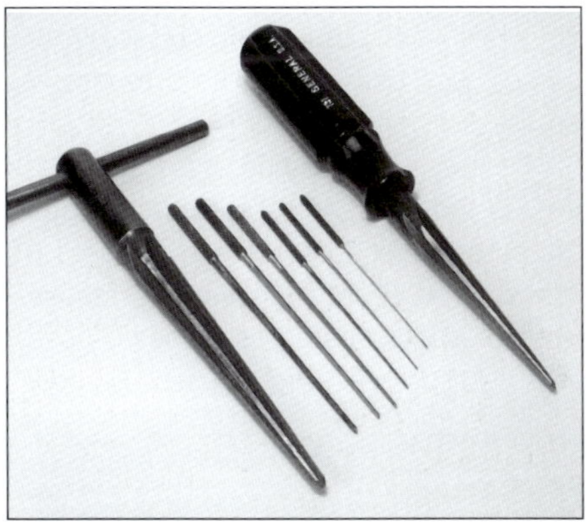

Reamers are extremely handy. The large T-handled job makes holes up to 1/2″ diameter; the plastic-handled version goes up to 3/8″. The set of six micro-reamers is perfect for model work; it comes from Micro-Mark.

cate — if you momentarily relax your grip on the pin vise, its weight may be enough to snap the bit. For these tiny drills I use a miniature pin vise that's so light the drill can support it.

Before drilling, make a dimple to start the hole so the drill won't wander across the plastic. A heavy darning needle in a pin vise or dowel handle is ideal for this.

Sharp drills cut styrene easily with very little pressure, producing a spiral chip. Remove the bit after every two or three turns to clear the chip so it won't jam and break the bit. If the drill seems hard to turn, try lubricating the tip by twisting it in a cake of beeswax. Oil can be used as a lubricant, but it may discolor the plastic and its residue must be washed away after use.

Reamers are useful when you need to slowly enlarge a drilled hole for a precise fit, or when you don't have the exact size drill you need. The tiny reamer set offered by Micro-Mark ranks as one of the most-used tools on my workbench.

Detailing with paint

Gluing bits to the model is not the only way to achieve a detailed look. Picking out parts with contrasting colors adds visual interest to the model, and is every bit as good as three-dimensional detail, sometimes better.

A perfect example of this is painting the engraved detail of the instrument panels provided in most aircraft kits. Painting the instrument faces white or black, dry-brushing them, and flowing on clear gloss lenses makes the panel look as if you spent hours — not minutes — detailing it.

Detailing projects and tips

In the sidebars of this chapter several talented modelers demonstrate techniques you can use to detail your models. They won't help you decide what details to add or where to stop, but they'll show you how it's done.

REPOSITIONING AIRCRAFT CONTROL SURFACES

by Pat Hawkey

When airplanes are shut down and parked, various control surfaces, either through loss of hydraulic pressure or simple gravity, droop out of the neutral positions that models are nearly always molded in.

Separating flaps, elevators, and rudders and repositioning them adds significant life to otherwise static-looking aircraft. To accurately represent moved control surfaces requires more than just cutting them off and gluing them back on at a different angle.

The Hobbycraft F7U kit includes separate ailerons, but as they come, **A**, they won't accurately depict anything but the neutral position. The large gap between the trailing edge of the wing and the leading edge of the aileron would not be there on the real airplane.

The first step is to carve away the

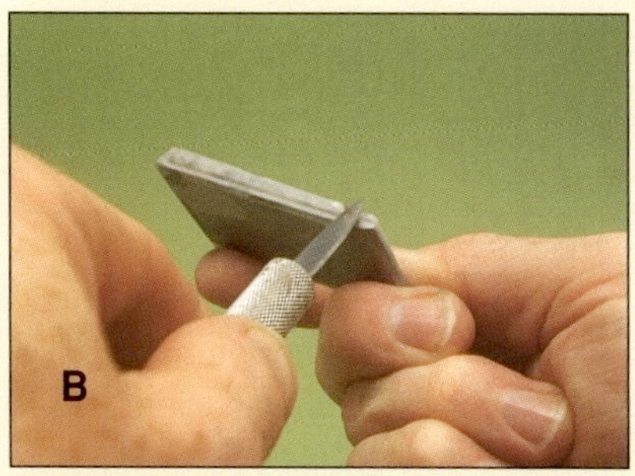

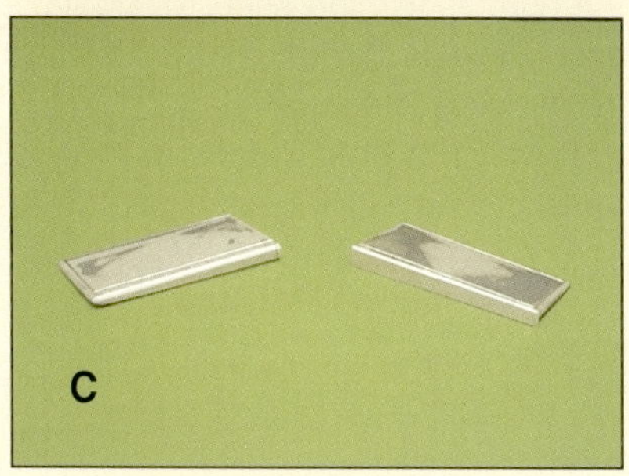

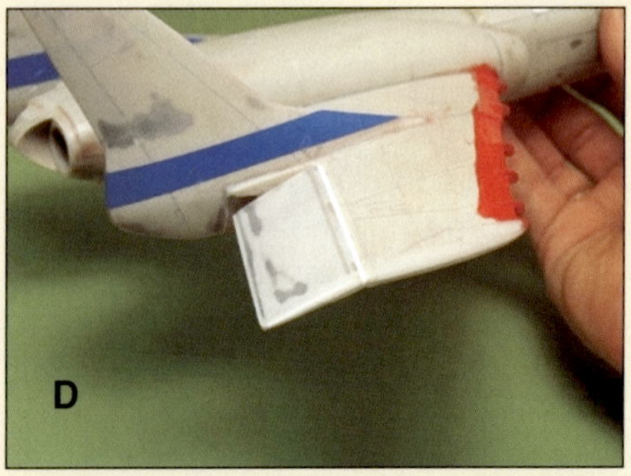

D

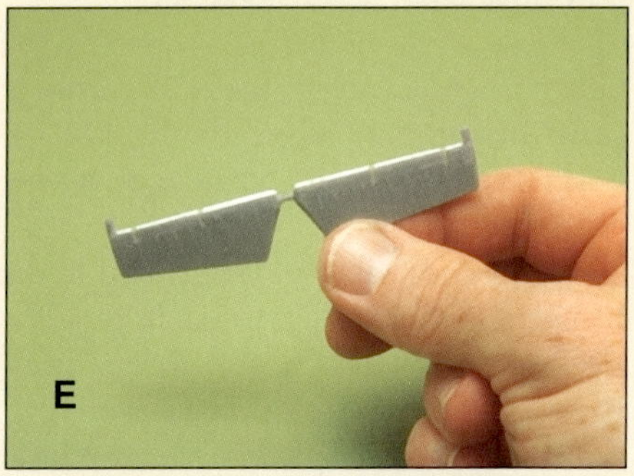

E

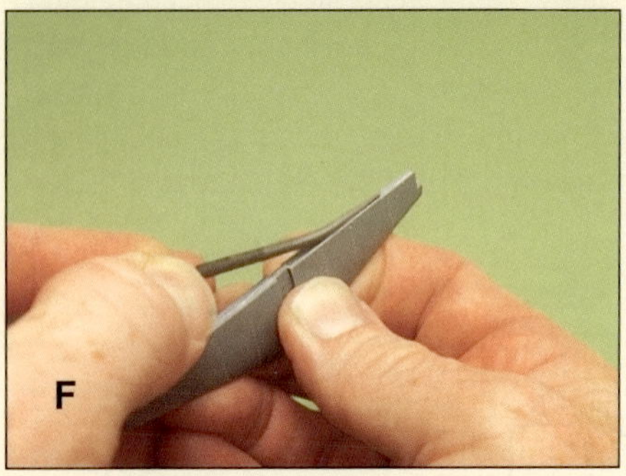

F

G

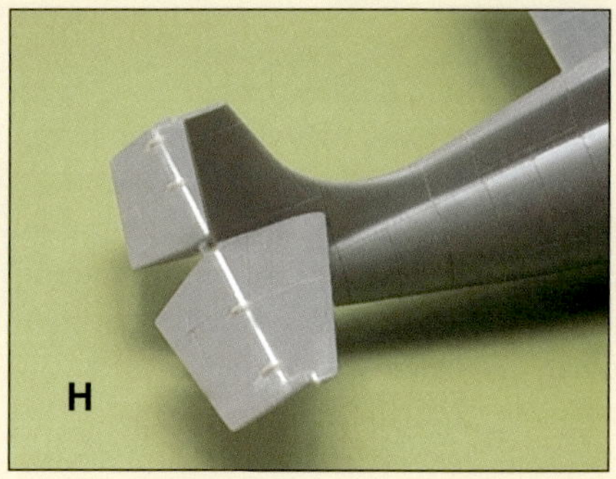

H

raised portion molded into the aileron, **B**. A strip of .040″ sheet styrene is glued to the leading edge of each aileron, **C**. The piece on the left shows this. The piece on the right shows the sheet stock sanded until rounded and tapered to fit into the trailing edge of the wing.

Now the aileron droops as it should, and the sheet stock fills the gap that was there before, **D**.

(Top photo) The elevators on this Tamiya F4F were carefully cut away from the horizontal stabilizer, separating the flying and control surfaces but leaving two flat mating edges, **E**. Like the F7U ailerons, the leading edge of the Wildcat elevators were built up with a styrene strip, then sanded round and smooth.

To accommodate the addition to the elevators, the corresponding stabilizer surface was routed out with a round riffler file, **F**.

Further complicating this particular subject were the external hinges Grumman designed to make the elevators work. After carving away the Tamiya renditions, I fabricated new ones. These were .010″ rectangles with holes punched out of them with a punch-and-die tool, then the edges around the holes were rounded. The four hinges were set into grooves sawed into the stabilizer and secured in place with a swipe of liquid cement, **G**.

After careful filing, sanding and fitting, the desired effect, **H**, was achieved

An airbrush is the paint applicator of choice for styrene models. Note the paper towel tube in use as a painting fixture, and the disposable plastic glove on the hand holding the model.

Chapter 8

Basic Painting and Finishing

PAINT CAN MAKE OR BREAK a model. No matter how fastidious your fabrication, assembly, and detailing, what you see on the completed model is the paint job. While I've seen a great paint job rescue a so-so model, I've seen a lot more cases where a beautiful construction job was rendered mediocre by poor finishing.

Because painting is the finishing step in every modeling project, it's an absolutely essential skill. Many moons ago I purchased my first air-

Scrub-a-Dub-Dub! The main step in preparing a model for painting is cleaning it. This styrene kit structure is getting a bath in warm water and dishwashing detergent. The toothbrush removes dust from recessed detail areas.

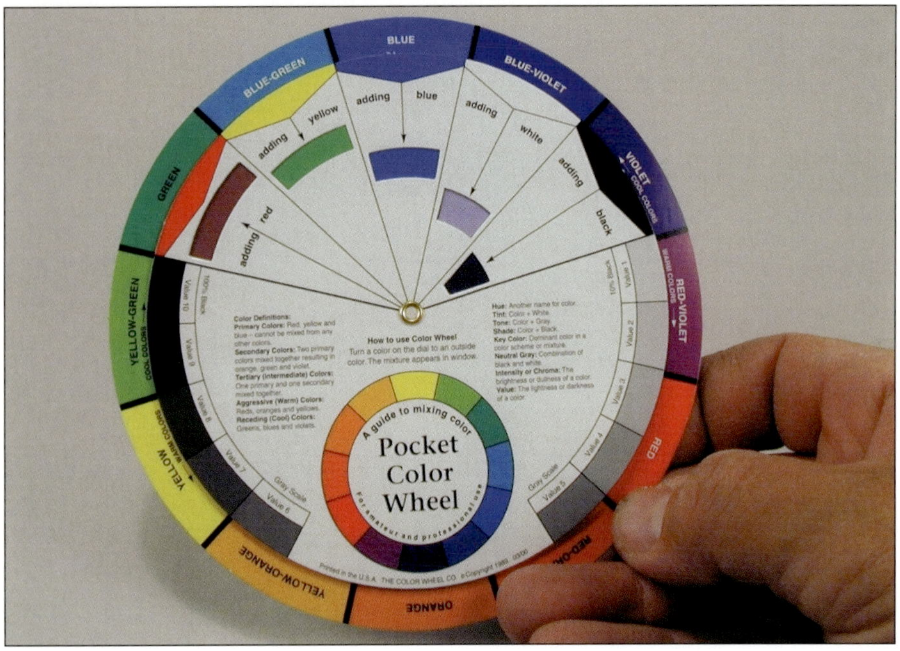

This Pocket Color Wheel has a rotating middle section with windows that indicate what colors to mix to achieve the hue you want. The back has a similar rotating display that shows color complements and harmonies.

brush and taught myself to use it by trial and error (a lot of error!). Since then, I've settled on the airbrush as the best way to paint almost everything I build.

For models, the paint coating needs to be thin — probably not exactly to scale, but thin enough not to obscure detail, and skin-tight. Virtually all of today's model paints incorporate fine-ground pigments to achieve this.

First, cleanliness

Styrene models must be cleaned before painting. This isn't just whisking away the big pieces, it's a thorough cleaning with detergent and water, gentle solvents, or both. I wash styrene models in warm water and dishwashing detergent, then give them a final dip in clean denatured alcohol.

After the model is clean and dry, handle it only with disposable rubber gloves, and cover it to keep it dust-free until you apply the paint – preferably right away.

Mixing colors

Model paints come in a bewildering variety of colors, so you won't often have to mix your own. When you do, start with the lightest colors and add the darker ones until you have the shade you want. Stir frequently, and let the paint dry on a piece of white cardboard or white sheet styrene so you can assess the color accurately.

Color mixing is something of an art, and you'll find a color wheel at art and craft stores to help you. The general rule is that you should combine the fewest possible colors to achieve what you need for the model. This keeps the color from becoming muddy.

Most often, you'll mix paints to achieve scale color. This concept is best summed up by the equation:

Distance = dulling

The further you stand from an object in the real world, the duller and grayer it appears. Since our viewing distance for models does not vary appreciably with their scale, this means that the smaller the scale, the duller the colors should be. Most modelers achieve this scale effect by mixing full-strength colors with varying amounts of white, light gray, or gray-blue.

Keep a record of the special mixes you make, particularly if you know you'll need the same color again. A good way to do this is to spray the color on a scrap of styrene or white

PAINTS FOR STYRENE MODELS

Virtually all of the paints offered specifically for model work are high-quality coating systems, and all of them can be made to work on styrene. Paint comes down to choices: What brands have the colors you need? How will you apply them? And finally, What are you comfortable with?

The most important thing about choosing a paint system is that you choose. The alternative is to have half a dozen colors from each of half a dozen brands, none of which can be mixed. This is OK while you're experimenting, but once you find a brand of paint that you can achieve consistently good results with, stick with it.

Acrylics

Water-based and water-cleanup modeling paints have been available for a couple of decades. The brands include Testor Acrylic, Polly Scale, Tamiya, and Accu-Flex, and it seems as if a new line comes along every couple of years.

Acrylics are easy to use, and especially easy to clean up. Properly handled — which usually means simply following the manufacturer's instructions — they airbrush beautifully, dry quickly, and adhere well.

Acrylics are fussy about cleanliness, and any traces of finger oils or lubricants from molding machinery on the model can repel the paint — oil and water don't mix! Thoroughly cleaning the model with water and detergent will prevent this.

Other than cleanliness, the main concern with acrylics is getting them to flow smoothly. This is because the surface tension of water is much higher than mineral spirits or other petroleum-based solvents.

The solution — literally — is to add a wetting agent to the paint at the same time you dilute it. The home remedy is a couple of drops of liquid dishwashing detergent in a pint of water; commercial preparations include Liquitex Flow-Aid, Kodak Photo-Flo, and airbrush thinners from the makers of the paints.

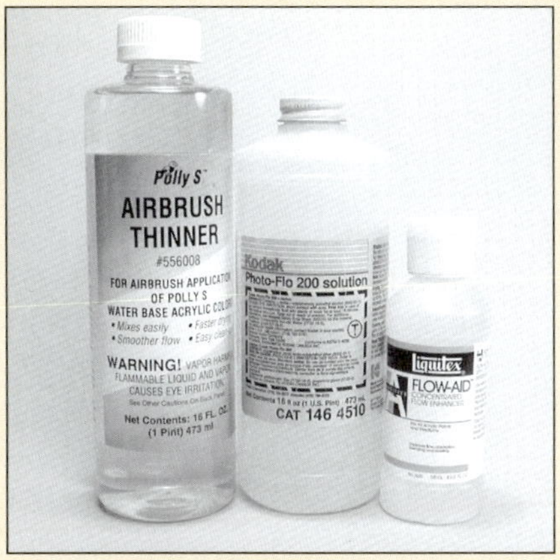

Acrylic water-based paints need help to get them to flow smoothly on a styrene surface. A drop or two of one of these wetting agents will do the job.

All of them use detergent, alcohol, or both to make the paint "wetter."

One note of caution: Acrylics are airbrush killers. Leaving even small amounts of acrylic paint residue in your airbrush can all but weld it shut, and once dry, acrylics are impervious to water. Since the cleaning solvent is water — maybe with a little denatured alcohol or ammonia added to make it clean better — there's no excuse for not cleaning your airbrush thoroughly.

Enamels

Enamels are the old standbys for painting plastic models. The most widely available brands are Testor Model Master, Humbrol, and Floquil. (Floquil's thinner, Dio-Sol, includes xylene, which crazes styrene badly. For that reason Floquil paints should only be applied to styrene with an airbrush, and then with extra care.)

Most enamels are thinned with mineral spirits (or their own proprietary thinners), and can be cleaned up with lacquer thinner. These are petroleum-based (organic) solvents, and must be used with good ventilation — for airbrushing, an exhaust booth is all but mandatory.

Enamels are easy to airbrush, and they lay down smoothly when properly diluted. Properly applied, they become impervious to their own solvent after curing for several days.

Lacquers

Lacquers are one of the oldest types of finishes. Most of the model lacquers are intended for use on metal or wood, and are poor choices for styrene. Some of the brands offer a barrier coat, but it is rarely worth the extra trouble.

True lacquers have the properly of being dissolvable by their solvent forever, even years after application. For styrene this is actually a significant disadvantage, since we usually want the base coat of paint to protect the plastic surface from crazing during subsequent finishing steps.

— Bob Hayden

Here's a sampling of the water-based acrylic colors available for painting styrene. The Accu-Flex and Tamiya colors (left and center) each have their own thinners; the Polly Scale Military and Railroad Colors (right) thin with distilled water.

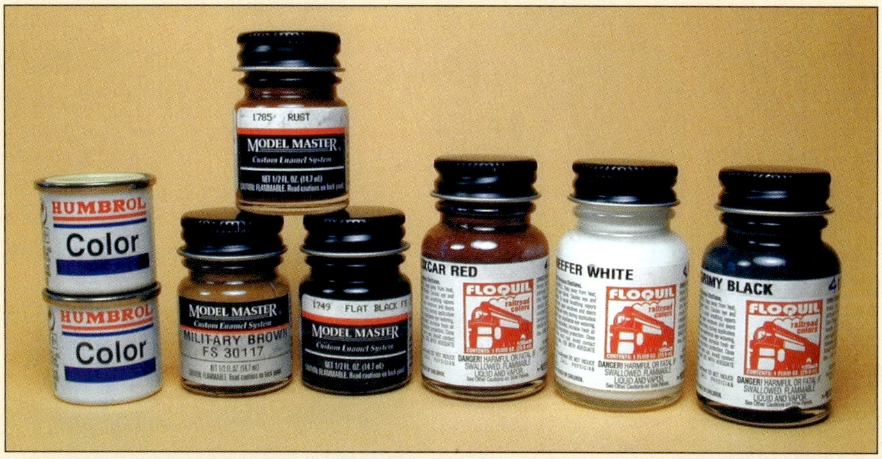

Enamels are the traditional paints for finishing styrene. The broad range of Humbrol colors (left) comes from England. Testor Model Master enamels (center) are widely available in hobby and craft stores. Floquil Railroad Colors (right) have a strong solvent, and must be airbrushed carefully on styrene models.

Modelbuilders will argue vociferously about their favorite clear finishes. From left to right: Testor Dullcote, Testor Model Master Lusterless Lacquer Overcoat, Floquil Glaze (a glossy additive to make the paint stick when thinned), Floquil Clear Flat, and Floquil Crystal-Cote (clear gloss).

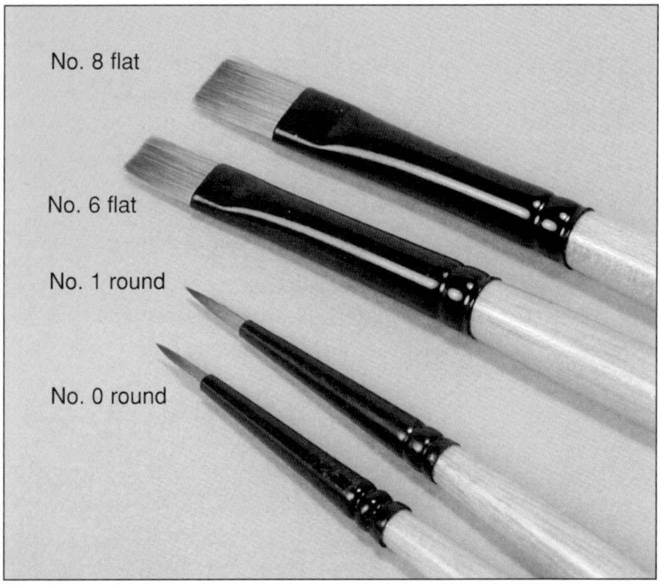

These high-quality red sable artist's brushes are ideal for model work. The broad-tipped flat brushes at top are good for applying smooth base coats; the smaller rounds at the bottom are what you'll need for picking out details

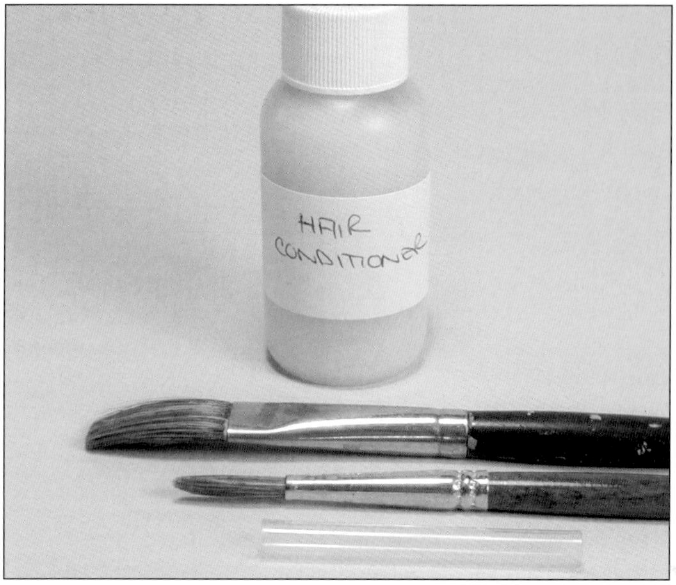

The two brushes in the foreground have been used regularly for more than a decade. One reason for their longevity is regular applications of hair conditioner after cleaning. The clear tube next to the round brush also helps preserve its bristles.

cardboard at the same time you spray the model. Then record the mixture specifications on the back of the sample and put it away for future reference.

Do not mix different brands of paint, and do not thin model paints with anything but what the manufacturer specifies. In addition to solvents, many proprietary thinners include retarders or conditioning agents to make the paint perform as specified. While you can often substitute a cheap solvent such as lacquer thinner for cleanup, don't add it to the paint.

Acrylic paints are especially fussy about thinners. Some brands may be thinned only with water, and even a trace of alcohol will turn them into a gooey mess. Read the instructions!

Most brands caution against returning thinned paint to the bottle. This is especially true with acrylics, where the manufacturers recommend discarding thinned colors if they will not be applied within a few days of mixing.

Enamels and lacquers can be safely returned to the bottle after thinning. Add a BB to the bottle to remind you that it is no longer full-strength paint. Otherwise you may forget and overthin the paint the next time you apply the color.

Brush painting

Brushes are wonderful for applying weathering treatments and painting small details, but it's difficult to obtain a smooth, thin base coat of paint with them. Brush painting is also time-consuming compared to airbrushing or spray-can application.

High-quality red sable artists' brushes are what you need, and the prices can be absolutely breathtaking. Depending on the size of the model and the surfaces you'll have to cover, you'll need a couple of No. 6 or No. 8 flat-tipped brushes for broad areas and two or three No. 1 or No. 0 round ones for picking out details.

Don't be fooled into buying the very small brush sizes — No. 000, 0000, and finer. These hold almost no paint, and their work is better done with a sharp-point on a No. 0.

The keys to brushing smooth base coats are to moisten the brush with thinner — either water or solvent — and to adjust the consistency of the paint before applying it to the model. The first is easy: Just dip the brush into the thinner and squeeze out the excess, leaving the brush hairs damp but not wet.

Thoroughly mix the paint before application. Some of the newer water-thinned model paints caution against shaking the bottle (it causes bubbles), so check the instructions and stir if that's what they specify.

Right out of the bottle the paint is usually too thick to flow smoothly. Thin it before brushing it on the model, either by adding thinner to each brush load, or by thinning a supply in a separate container.

Make a styrene palette from a scrap of sheet stock and use it to check the paint consistency. Dip the brush hairs halfway into the mixed paint, then spread the color across the palette to work it into the hairs. This will let you see and feel how the paint is going to flow onto the model.

Brush the paint on the model in short strokes, going in one direction and never going back over the same area twice. If the model needs more paint for coverage, apply a second coat after the first has dried. Otherwise you'll have brush marks for sure. Most jobs will require two coats.

For small details, or for freehand painting of camouflage patterns, reach for the No. 1 brush. Form it into a sharp point with your fingers or by rolling it in paint on your styrene palette, then use the point or side to apply the paint. If the area is large, paint the edges with the fine brush, then switch to a flat brush to fill in the center.

CHOOSING AN AIRBRUSH

Airbrushing is the best way to apply base color coats to styrene models, as well as some special effects. While spray cans and automotive touch-up guns come in useful for very large projects, the airbrush is the all-around miniature paint sprayer.

For modeling, the airbrush is not an artist's tool, but a miniature paint gun. Artists typically spray very thin paints and dyes onto paper, which is a whole different ball game from spraying our heavy pigmented modeling paints. Because the instructions are usually written for artists, it's no wonder that a lot of modelers have trouble with this tool.

Single- or double-action?

There are several kinds of airbrushes. They're classified by whether the control button governs only the flow of air (single action) or the flow of air and the flow of paint (double-action). On double-action brushes you push down for air, and pull back for paint. This allows you to vary the paint volume continuously, which requires additional coordination and makes double-action brushes harder to learn.

Internal- or external-mix?

Airbrushes are also classified by where they mix the air and paint. Internal-mix models bring the paint inside the body of the airbrush; in external-mix units the atomization of the paint takes place outside the body.

Relatively simple, mid-range airbrushes work best for me, and I recommend a single-action, external mix unit for your first airbrush. Such an airbrush is easy to learn, easy to use, and easy to clean.

Medium tip

For airbrushing model paints you'll usually want the medium tip, needle, or paint control. Compared to the inks and watercolors that most airbrush artists use, our paints are more like spraying glues — an important point to remember when cleaning the airbrush after use.

Airbrushes don't have a lot of moving parts, but the passage of the air, paint, and solvents through them and repeated cleanings wear them out. Replacement parts are available for most brands.

Color cup or bottle?

Most airbrushes come with open-top color cups to hold the paint. This is fine for artists, where the airbrush is pretty much held straight and level for working on an easel (or a T-shirt).

I greatly prefer color bottles, the kind with a lid. Painting models usually involves tilting and tipping the airbrush, and whenever I use a color cup I wind up with paint running down my arm and dripping off my elbow.

Although the bottles have an air vent where the paint can run out, the cap keeps most of the paint inside. Bottles also have a siphon tube to which you can attach a wire-mesh strainer to filter out paint clumps. — *Bob Hayden*

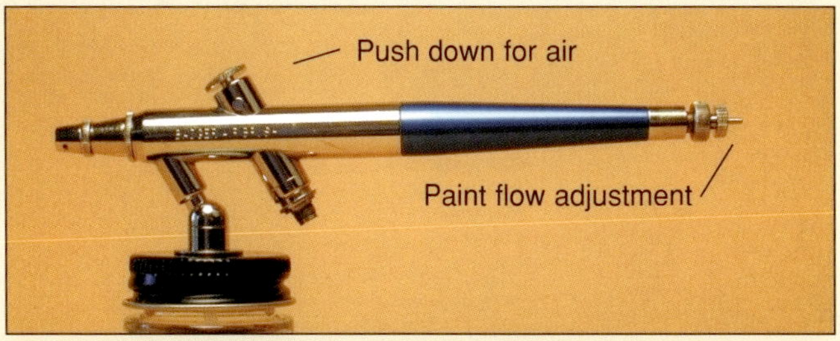

This Badger Model 200 is a typical single-action, internal-mix airbrush. A long, tapered needle inside the body regulates the paint flow, and is adjusted with the knurled nut at the rear.

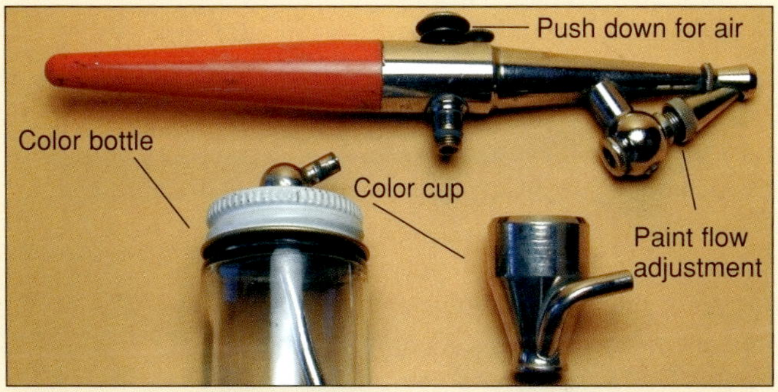

My well-worn Paasche H model, a sturdy single-action, external-mix airbrush. Paint flows into the cone-shaped color control and is mixed with air in the nozzle, outside the airbrush body. It has a No. 3 (medium) tip, ideal for model paints.

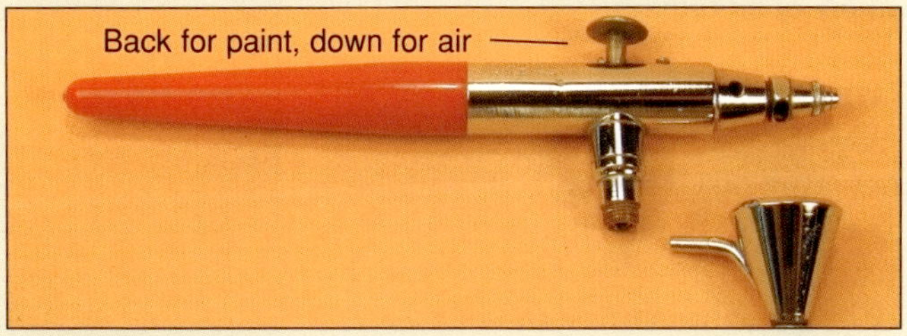

This Paasche V2 is a double-action, internal-mix unit more suited to artwork and retouching than model painting. The control button goes down for air and back for paint, and the paint volume can be varied continuously.

This Testor Aztek airbrush can function as either a single- or double-action machine, and the interchangeable paint tips (right) offer a broad range of spray patterns. They come out of the airbrush body for cleaning. The design also accepts either color cups (center) or color bottles (left).

This battered warrior, a Binks diaphragm compressor, is 30 years old and still chugging away. An entry-level air supply for airbrushing, it's relatively quiet and will get the job done.

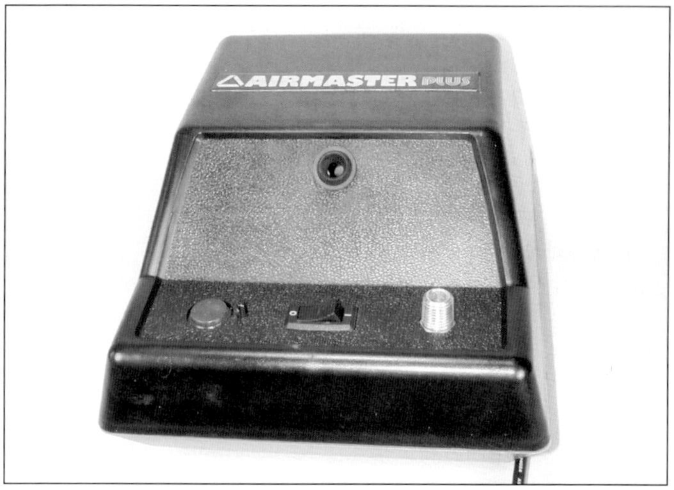

Originally intended for the cake-decorating trade, this compact Airmaster compressor is quiet and demand-regulated: It only starts up when you depress the air button on your airbrush.

Although it makes enough noise to wake the dead, this 3/4-h.p. home-shop compressor is an excellent source for airbrushing. The motor charges the tank, then cuts off until the tank pressure falls below a preset level. A quick-disconnect fitting on the outlet pipe allows swapping the airbrush and its regulator for a variety of air tools.

Caring for quality brushes

Good brushes are expensive, but they can last for years if you take good care of them. Follow these three steps:

Clean the brush first in your everyday dirty thinner, then rinse again in clean thinner. This removes most of the paint from the brush hairs. Never scrub the hairs against anything harder than a wet bar of soap.

Wash the brush with warm water and soap (you can buy commercial artist's brush soaps, or use plain shampoo). Rinse with clear water, then gently dry the brush with a rag or paper towel.

Finish by working a drop of hair conditioner into the hairs and mold them to the brush's original shape. Then store your brushes upright in a coffee can or Styrofoam block so nothing can touch and deform the hairs. Keep the clear plastic tube protectors that came on the small brushes when you bought them. They're good for protecting the tips during storage.

This simple brush-care drill will take a couple of minutes per brush each time you use it, but you'll spend a lot less time and money buying new ones at the art supply store.

Spray cans

Aerosol spray cans are useful, especially for painting large models quickly, but they lack the precision and fineness of airbrush painting. They have only one spray setting — Full On — and they apply a lot of paint. This can lead to runs and drips on the model, and a great deal of wasted overspray.

If you must use spray cans, start by warming the can in warm (not hot) water. This makes the paint flow better and level out quickly when it hits the model.

Start spraying to one side of the model, them move the can quickly across it and off to the other side. Never reverse direction with the spray pattern on the model — you'll get a double dose of paint and a run or glob.

Airbrushing equipment

The sidebar on page 61 will help you choose an airbrush if you don't already own one. In addition, you'll need a source of pressurized gas and some other accessories. Good news: All of these items cost a lot less than they did only a few years ago.

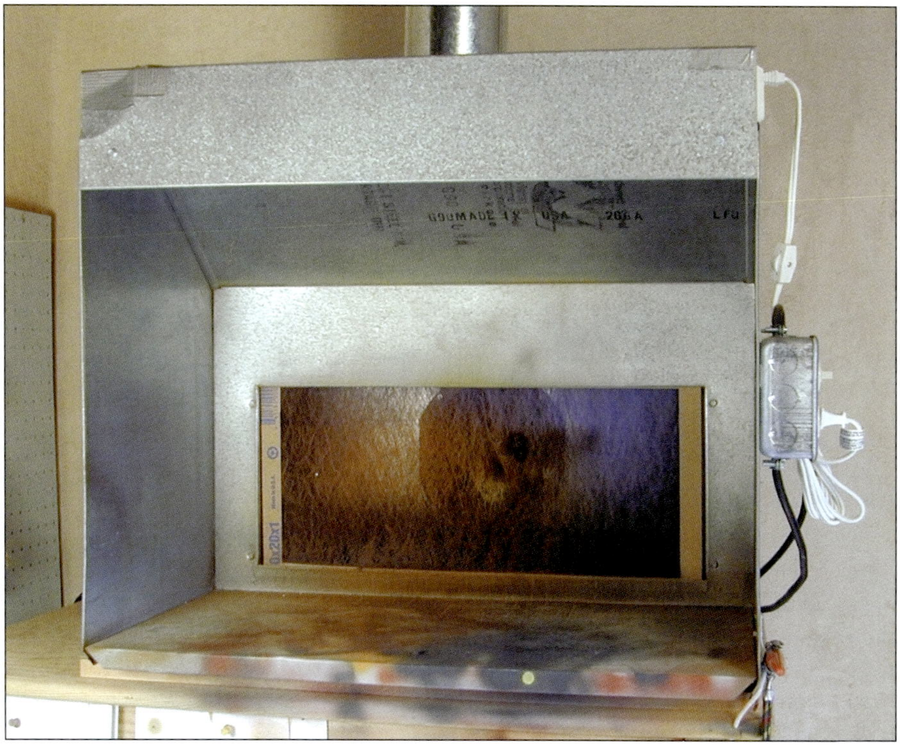

This North Coast Hobbies spray booth is the definition of "adequate ventilation." Its powerful blower exhausts solvent fumes out the 5″-diameter duct in the back, keeping them out of your lungs. Paint solids are captured by the replaceable furnace filter in the back wall.

Compressed gas

While there are alternatives such as canned propellants and rechargeable gas tanks, nothing beats the convenience or long-term economy of an electric air compressor.

As with airbrushes, there are several types and choices. For years I used a Binks diaphragm compressor, which was relatively quiet but suffered from pulsations in the air pressure at the airbrush. Adding a 12-foot hose between the compressor and the airbrush helped, and the problem was solved when I added an inexpensive 5-gallon portable air tank between the compressor and regulator. The tank served both as a surge suppresser and an efficient moisture separator.

Next I tried a Campbell-Hausfield 3/4-h.p. tank-mounted general-purpose shop compressor. The rig includes a pressure gauge and regulator, and shuts down when the tank is charged up. It even has wheels and a handle to make it easy to move around. Its sole disadvantage is that it makes more than enough noise to wake the dead.

If noise is a major consideration, look into the noiseless compressors available from art-supply houses. Similar to the silent compressors used in refrigerators, they are truly close to noiseless — but expensive. They're surely worth it for apartment dwellers, or if you work far into the night while the rest of the family sleeps.

Yet another choice are the quiet, demand-operated compressors sold for crafts and cake decorating. These are compact, inexpensive, and run only when you depress the airbrush button.

Regulators and moisture traps

The output pressure of the compressor needs to be throttled down to between 8 and 30 psi for airbrushing model paints. On my shop compressor I use the regulator on the compressor to reduce the pressure to 60 psi or so, then use a Norgren air-

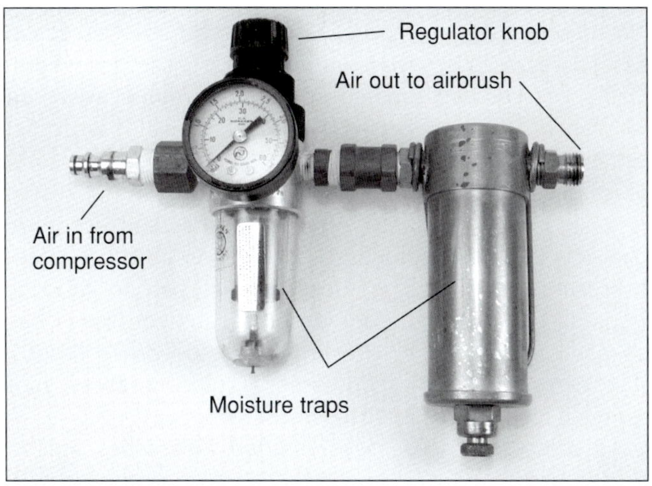

Here's a belt-plus-suspenders approach to moisture traps. The device with the gauge reduces the output pressure of the compressor to what the airbrush can use. The can-shaped metal item is a second moisture trap, used to ensure dry air output during humid summers in Milwaukee, Wisconsin.

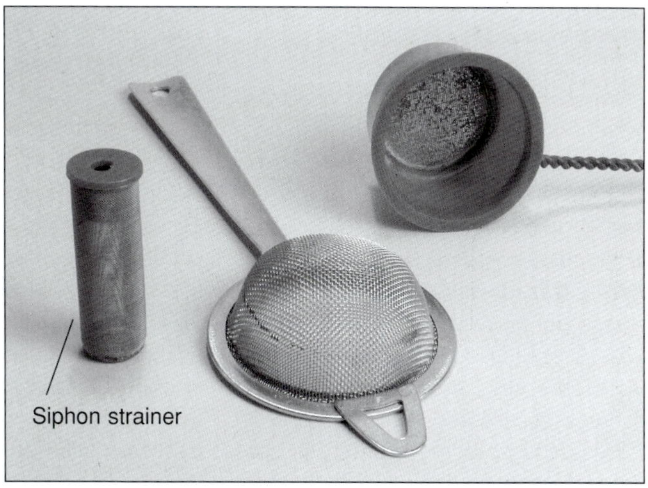

Straining lumps out of the paint is one of the keys to successful airbrushed finishes. The small tea (center) and enamel (right) strainers are used before you airbrush. The cylindrical mesh strainer on the left slips over the siphon tube in the airbrush color bottle.

brush regulator-filter-trap to bring it down to 15-25 psi.

Most regulators include a dust and moisture trap. Getting the moisture out of the air supply is important. Otherwise, under humid conditions your airbrush can spit a big gob of water on the model along with the paint. This does not make for top-notch finishes.

PAINTING *after* ASSEMBLY

by Jack Burgess

Unless the finished model will be a single color, there are generally two approaches to painting models. One discipline advocates painting all subassemblies prior to final assembly. Using this approach for a building for example, all windows, doors, corner trim, and other items which are to be painted the trim color are painted before assembly. Then, once the building sides are assembled, only touch-up is needed.

The second approach (mine) postpones painting until the building is complete. Although it can result in a lot of masking, I prefer this latter technique. Being relatively impatient, I like to see a model come together as quickly as possible; I don't like to wait for individual subassemblies to be painted before they are added to the model.

There are benefits to painting after assembly. I don't need to be as careful when bonding things together — there is no need to worry about cement ruining an already-painted surface. While some prepainted items such as windows and doors can be glued in place from the inside, it can be difficult to glue things like prepainted trim to a building without marring the adjacent finish.

Painting after construction also makes it easy to discard something when things don't go as planned. I might have second thoughts about discarding a major assembly that has already been painted, but I don't have a problem tossing an unpainted piece and starting over.

The Standard Oil Office at El Portal, California, still exists near Yosemite National Park. This HO (1/87 scale) model was scratchbuilt from scribed styrene. Jack Burgess filled the hollow roof and ceiling on the model with casting plaster to eliminate any possibility of warping, and painted the model after completion.

I occasionally delay installing windows and doors until the building is painted its base color. Then it's a simple matter to paint the windows at the same time as the trim and install them in the building when everything is dry.

But most of the time I finish an entire structure before starting any painting. The base color is first airbrushed onto the entire model. Once this dries, I mask off everything except trim items with Scotch Magic Mending tape and airbrush the trim color.

I always paint styrene models with an airbrush, saving brushes for figures and detail painting. I have found that paint must go on relatively thick to adequately cover when brush painting. It can also be difficult to avoid brush marks.

As for paints, I have used Floquil Railroad Colors for 30 years and continue to use them even though water-based paints are now available. (I have a powerful paint booth to exhaust the solvent fumes.)

Sheet styrene is much more resistant to Floquil than injection-molded plastic, and I have never found a need to apply a barrier coat to a model before airbrushing a Floquil color coat. I airbrush in light passes, especially on plastic castings such as doors and windows. This avoids problems of the paint etching the plastic.

Adequate ventilation

Spray painting is one of the few modeling operations where you can seriously hurt yourself. Solvent fumes in the air are not good for you to breathe, and even the water-thinned paints have bad things in the paint solids that you don't want in your lungs.

If you only paint a few models a year, "adequate ventilation" might be painting outdoors, or in your garage with the door open. If you can still smell the solvent in the paint, you can achieve a measure of protection if you wear a respirator mask with replaceable cartridges that absorb organic compounds. You can find these at large hardware stores or farm centers, and the cost is reasonable.

If you paint frequently, nothing beats a powerful exhaust booth, ducted outdoors. Such an appliance may cost several hundred dollars, but I count mine as one of the smartest hobby purchases I ever made. There is no greater luxury than walking over to the spray booth, switching it on, spraying away, and knowing that your health is protected.

Also on the safety front, wear at least one disposable latex examination glove whenever you spray paint. The glove allows you to hold the model, and it means you won't have to douse that hand with solvent or spend time scrubbing away paint residue.

Most loaded airbrushes will fall over if you set them down on a flat surface, so you need a holder for them in your painting area. You can buy one at your art supply store, or you can make up something simple by bending coat hanger wire. You'll be amazed how much having a secure place to set the airbrush simplifies painting with it.

How to airbrush

Airbrushing is not a difficult skill to master if you're willing to buy the right equipment, follow directions, and spend some time practicing. (Sounds like learning to play the piano!)

There are three variables to deal with: How thin the paint is (or needs to be), the volume of paint the airbrush is set to deliver, and the air pressure being delivered to the airbrush.

While one of these variables may change slightly from color to color (light colors, for example, have more pigment per drop of paint than dark colors, and may require a little extra air pressure), for all intents and purposes once you've found a combination of settings that works, you're in business.

Paint dilution and straining

Even paints that claim they can be airbrushed right from the bottle will need at least minor dilution. Thicker paints can call for as much as 50 percent thinner (one part paint, one part thinner).

Start with the manufacturer's recommendation for thinning. In the absence of instructions, use two parts paint to one part thinner as your starting point. If the paint won't come out of the airbrush, add more thinner; if it sprays on easily but looks watery and won't cover, use less.

The only real trick I've found to airbrushing is always to strain the paint. You can use a fine-mesh tea strainer when you decant the paint into the color bottle, or an in-bottle wire mesh strainer that fits on the siphon tube. You will be surprised at how many clumps come out of a well-mixed bottle of brand new paint, and each is a clogged airbrush tip or needle and a ruined painting session just waiting to happen.

Spraying pressures

Begin your quest for the perfect airbrushing pressure by reading the airbrush and paint manufacturers' recommendations.

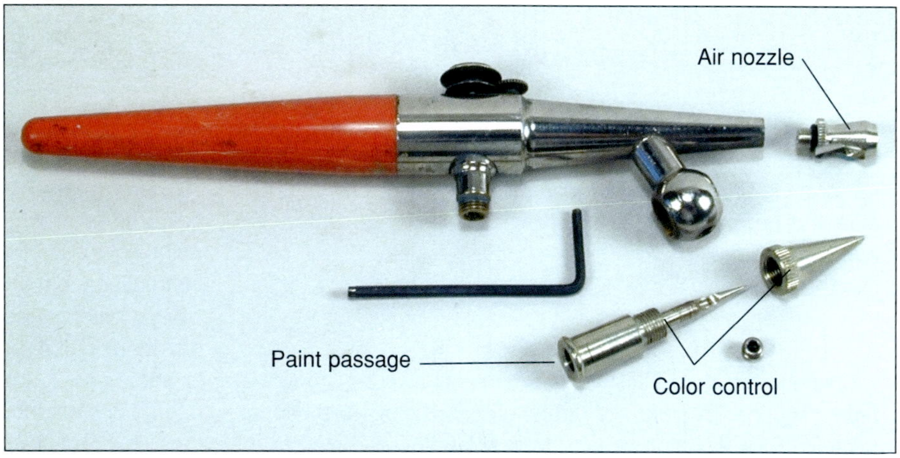

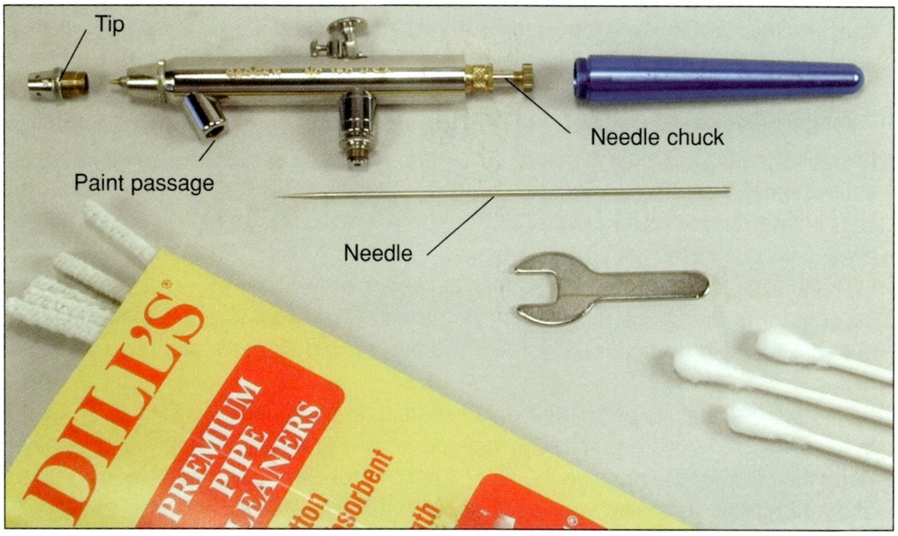

You don't have to completely field-strip your airbrush every time you use it, but frequent partial disassembly is required to ensure that critical paint and air passages remain clear. The Paasche H3 model (top) is easy to clean because the air and paint are mixed outside the airbrush body. The Badger 150 (above) has a needle to control the paint flow, which should be removed and cleaned often.

Start with the pressure at 17-18 psi, then work up or down. Thin and strain the paint, load up the airbrush, and try various volume settings on a sheet of clean white styrene or an old model. Evaluate the results, then raise or lower the pressure about three pounds and test again.

Water-base acrylic colors require an extra pound or two, but not much more than that. Your average pressure should fall somewhere between 8 and 24 psi — anything higher than 30 psi means something is wrong: The paint isn't thin enough, the color control isn't providing enough paint, or the needle/tip combination is too fine for the relatively heavy paints we spray on models.

Move the airbrush smoothly across the model, with the tip from 2" to 8" from the surface being painted. This is closer than most instructions call for, but as long as the paint glistens briefly when it hits the surface, then dries, it's correct. It's better to build up coverage slowly than to apply too much paint and cause a run or puddle.

Apply paint from inside to out, spraying interior surfaces and inside corners, then broad exterior surfaces and outside corners. Stop every now and then to assess how you're doing — there's a tendency to be drawn into the work and apply too much paint.

A STANDARD AIRBRUSHING ROUTINE

Once you've achieved proficiency with your airbrush, it's helpful to settle on a standard routine for preparing, painting, and cleaning up afterward. The idea is to limit the variables and avoid problems.

1. Prepare the models and parts

Cleaning the airbrush is work, and to minimize the amount of time I spend cleaning, I rarely airbrush only one model. Instead, I try to prepare parts from two or more projects for spraying at the same time. One model may be ready for a final coat of clear flat finish, another for a primer or overall color coat, and a third may have window and door or detail castings ready for painting.

Finding a convenient way to hold a completed model or subassembly during painting is important, since you don't want your airbrush session to turn into a juggling act. A bent wire coat hanger, paper towel tube, stick of wood, or other throwaway item makes the best painting handle.

The all-time favorite is probably the coat hanger, since it can be bent to accommodate anything, and you can even work it out so it will hold the model upright so you can set it down to dry.

Clean all the items to be painted, dry them, and arrange them by color on sticky-side-up masking tape on scrap cardboard. Write the color for the parts on the face of each cardboard flat, and arrange the flats near your paint booth in the order to be sprayed: clear coatings first, then light colors, dark colors, and metallics last. This sequence makes it easiest to clean the airbrush between colors, and reduces the problem of color contamination.

2. Prepare the paint

Paint needs to be thinned for airbrushing, and strained. Regardless of what a friend has told you about spraying the paint right out of the bottle, the best advice is to follow the instructions that the paint manufacturer provides.

Most paints need to be thinned from 15 to 50 percent for airbrushing — that's anywhere from 6 parts paint to 1 part thinner to equal parts! Light colors often need more thinner than dark ones, because they have more and heavier pigment to ensure their covering power.

Thin the paint through a tea strainer or other fine screen, or use one of the special in-the-bottle strainers made to slip on the airbrush siphon. Since I started straining all paint over 20 years ago, I've only had half a dozen clogs in hundreds of hours of airbrushing.

Arrange the prepared paints like the flats of parts, from clear to light to dark.

3. PAINT!

Crank up your compressor, adjust the pressure, and test the first color on a scrap model or piece of white cardboard. When you have a good smooth spray pattern, paint the first model or flat of parts. Run a little thinner through the airbrush between colors. Even after years of this routine I'm always amazed that applying the paint takes so little time compared to getting ready and cleaning up.

4. Clean up – thoroughly

Remove the painted models and paints from the spraying

Then set the model aside to dry in a dust-proof area. A large clean cardboard box turned upside down over the model achieves this nicely.

Give the paint as long as possible to dry before proceeding with additional colors or construction steps. I consider overnight drying a minimum, with two or three days even better. If you try to move on before the paint has set, you risk undoing earlier work.

Airbrushing "hot" paints

Paints with "hot" solvents will attack or craze styrene, just like the solvents we use for bonding. They can be applied to styrene using one of two techniques.

Most experienced modelers airbrush lacquers or enamels with strong solvents (such as Floquil Railroad and Military Colors) by misting on light coats that are almost dry when they hit the model. This takes practice, but as long as the paint dries immediately it won't have a chance to craze the surface. Once the base coat dries thoroughly, heavier subsequent coats can be applied without fear of damage.

These paint lines usually include a "surface prep" or barrier-coat product meant to be used for plastics. These can be brushed or sprayed, allowed to dry, and the standard paints applied over them. Such preparations are the only way you can brush paint lacquers over styrene.

Airbrush cleanup

Model paints, especially acrylics, are almost bulletproof after drying, and that means even small traces can weld your airbrush into a useless solid mass. For acrylic paints, spray lots and lots of water through the color tip, then spray some more! (Some brands suggest adding ammonia to the water to help with flushing. Others have proprietary airbrush cleaners. Use them.)

In my workshop the spray booth is right next to the laundry tub, so I can hold my airbrush under a stream of warm water for a couple of minutes while spraying to clean it. It's not much more work to haul a bucket of water to your spraying area.

Practice makes perfect!

The absolute best way to become proficient and confident with your airbrush is to paint models you don't care about. Use a half-finished model, an incomplete kit bought at a yard sale, or leftover parts from a kitbashing project to practice your technique

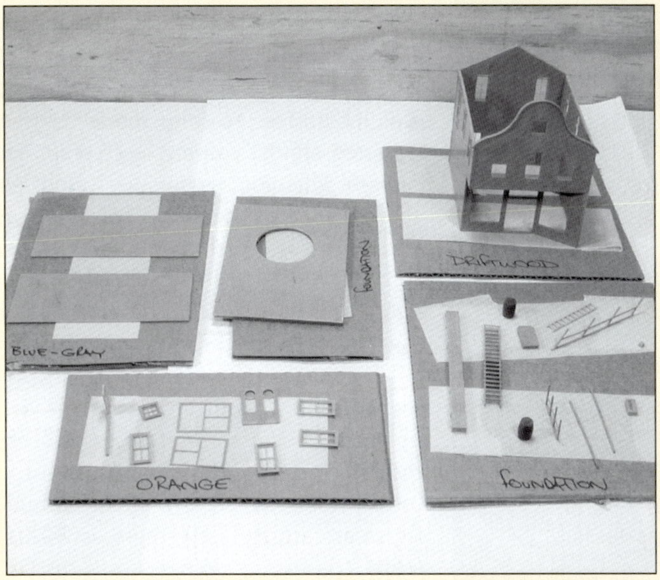

Setting up parts and subassemblies to be painted on scrap cardboard flats with sticky-side-up masking tape makes the models easy to handle during airbrushing. It also provides an easy way to remember the colors you plan to use.

The airbrush cleaners in the clear bottles are used to flush out acrylic water-based paints; the Dried Paint Solvent in the middle is what you use later, if you didn't do a good job of cleaning. The device on the right is a Testor Airbrush Cleaning Station. You load your airbrush with cleaning solution, then spray it into the fitting on top. Paint and solvent are caught in the replaceable white fiber wafer.

area, then thoroughly clean the airbrush. With my favorite Paasche H-3 external-mix model, cleaning requires only spraying solvent (water for acrylic colors) through the airbrush color control until it comes out clear, then gently swabbing the inside with a Q-Tip and a pipe cleaner.

Needle-type internal-mix airbrushes require some disassembly for cleaning, and you should always pull the needle to remove all traces of paint.

Finish by thoroughly cleaning all the gear that was touched by paint: color cups or siphons, strainers, eye-droppers, bottles, and so on. Cleanliness is indeed next to godliness when it comes to maintaining your airbrush and achieving good results with it, and the cleanup operation is actually the most important step in preparing for your next airbrushing session.

— *Bob Hayden*

without the pressure of obtaining presentable results.

As you practice, experiment with paint dilutions and spray pressures, but make sure you use the brand of paint you expect to choose for your models. (This is a great way to use up those two bottles of awful pink that you bought by mistake.)

Practice with dark colors on light models and vice versa, so you can see what the paint is doing, and be sure that you have plenty of lighting so you can evaluate and adjust the results as you go along.

Doug and Jackie Hole built this spectacular HO (1/87 scale) model of the Canadian National's massive steel arch bridge over the Fraser River in British Columbia. The 10′-long model is constructed almost entirely of styrene, and superb finishing techniques bring out the detail and exquisite construction. Photo by Don Weixl.

Chapter 9

Advanced Finishing Techniques

IF APPLYING A THIN, SMOOTH coat of paint to your model is important, what you do next is even more so. Adding markings, whether they are simple railroad lettering or colorful airline liveries, will bring the model to life. Altering colors and adding evidence of exposure to dirt and weather will give it a history.

These finishing techniques bring out the shapes, textures, and details on our models, just as variations in color and lighting make details more visible on the real thing.

Many modelers call these painting techniques "weathering," but it's instructive that some do not. Military modelers, especially the talented individuals who paint ultra-realistic figures, simply call their finishing steps "painting." Whatever term you like, these techniques are essential to producing realistic appearance.

Masking — it's not just tape

Complicated multicolor paint and camouflage schemes require masking, which is not difficult. Masking tape is only one means of preventing paint from hitting parts of the model, and many other materials can be used.

To use masking tape on models, first press it down on a sheet of glass and cut a fresh, sharp edge with a modeling knife or razor blade. Applying the tape to the glass also reduces its stickiness, or tack, which makes it less likely to pull up paint.

If you can find it, start with low-tack masking tape. Scotch Magic Tape, the frosty kind used at home and at the office, makes an excellent masking tape for models.

You'll also find fine-line masking tapes at the hobby shop. These are great for conforming to curves and complex surfaces. Use them to establish precise edges, then fill in broad areas with regular tape. Fine areas such as canopy framing can be masked completely over, then the areas to be painted cut out with a No. 11 blade and the tape peeled away.

Instead of masking them, cavities such as cockpits, fuselage interiors, or other places you don't want paint to reach can be stuffed with damp tissue while you paint. Testor offers Parafilm M, a clinging plastic film, as an alternative to tape for curved surfaces.

On extremely delicate surfaces such as natural-metal paint surfaces, use wet newspaper masks or the edges of 3M Post-It Notes for masking.

The masking principle I live by is that masking media should remain on the model for the absolute minimum time possible. Apply the masks, paint, and remove them as soon as the model is dry enough to handle. This avoids leaving tape residue on the model, and reduces the likelihood of lifting paint when you pull off the tape. Never leave masking on a model overnight.

Lettering and markings

Lettering can be painted on (a special category of masking) or applied using decals or dry transfers.

Dry transfers offer an advantage over decals in that they have no clear carrier film that must be hidden. They also have a huge disadvantage: Dry transfers cannot be prodded and nudged into precise position like decals. Once rubbed down, they're in position to stay, and the only way to adjust the fit is to remove them and try again.

Decals remain by far the most popular lettering and marking medium. (Some modelers even apply dry transfers to clear decal stock, so they can slide them into position!) The sidebar that begins on page 70 explains how to apply them.

Applying metallic finishes

Most models can be airbrushed without first applying a primer coat. One big exception is any model that will have a bare metallic finish. Metallic paints often don't cover well, especially over dark-colored styrene, and you'll need to apply either a white or light gray primer before the color coat.

Metallic finishes in general are fussy, because they will show up even microscopic flaws in the surface of the model. For that reason models that portray bare-metal subjects need to be almost mirror-smooth, and application of the finishing system can be nerve-wracking.

Bob Steinbrunn's 1/48 scale Seversky P-35 was built from a HobbyCraft Canada kit. In addition to extensive detailing, Bob gave the model a top-notch bare-metal finish. He used SnJ Spray Metal brand paint, but a main ingredient was preparation: Surfaces must be virtually glass-smooth for metallic paints and coatings.

The best advice is to choose a metallic paint or finishing system, follow the manufacturer's instructions to the letter, and practice with the coating materials before you apply them to a model that will be your pride and joy.

Modifying base colors and preparing models for weathering

We talked briefly about scale effect in Chapter 8. To simulate the effects of fading and to help make up for the vast disparity between the sunlight outdoors and the comparative gloom in which we view models indoors, I add substantial quantities of white to all the colors I use to paint models. This means adding anywhere from 5 to 25 percent white: 5 percent for light colors like tan and yellow, 25 percent for deep greens and blacks.

Nearly any type of weathering can be used on styrene once a base coat of paint has been applied, but any model that you set out to weather should have a cured, stable finish. This means a coat of clear flat finish

Text continues on page 72

Pastel chalks provide an easy and subtle means of adding dirt and rust to a model's surface. The left half of this HO (1/87 scale) hopper car shows the chalk powder as applied; the right side has been treated with Testor Dullcote spray.

WORKING WITH DECALS

Decals are magic! Even a handsome painted model comes to life when you add lettering and markings, and in the case of airliners and colorful combat aircraft, the effect can be even more dramatic.

It helps to understand how decals work as you learn to apply them. A decal consists of a clear film carrier with paint or ink markings printed on top of it. The clear film rests on a coating of one or more adhesive layers that hold it to an absorbent backing paper, and when softened with water, to the model.

On some decal papers the adhesive that holds the marking to the backing is fast-dissolving, and the adjacent layer dissolves slower. This gives you time to slide the decal into place before the adhesive washes away. For this reason, decals should be applied quickly — don't leave them floating in the water.

"Silvering," the number-one flaw in decal application, occurs when air is trapped between the decal carrier film and the surface of the model. All decal application techniques are aimed at eliminating it.

Setting agents

You'll find several brands of decal setting agents at your hobby shop, **A.** My favorites are Microscale Micro Set and Micro Sol. Micro Set is a wetting agent that contains diluted acetic acid; it has a faint vinegar smell. It does not soften the decal film like other setting agents, but helps eliminate silvering by helping the decal film slide smoothly across the model surface as you prod it into position.

Super Sol, Solvaset, Decal-Set, and other setting agents are solvents, usually diluted alcohol with other chemicals. These soften and dissolve the clear decal film and allow the markings to snuggle down against the model's surface, even when it has raised or recessed detail.

Setting agents vary in strength. If you find a setting agent too strong for a given brand of decals, dilute it with distilled water. Some decal brands call for their own proprietary solutions.

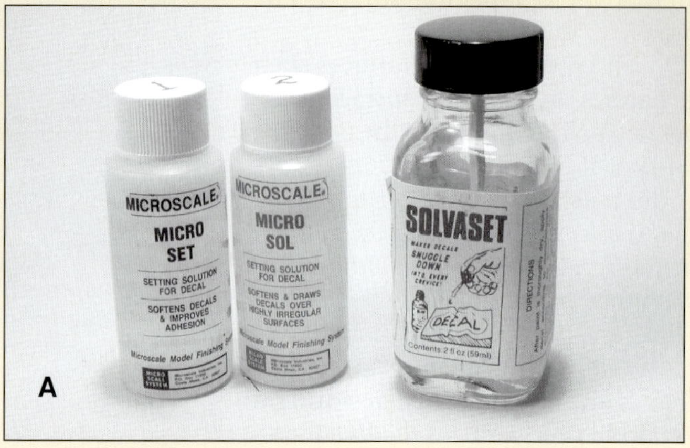

A

Step-by-step decal application

Decals must be applied to a smooth, glossy surface; trying to apply them over a flat finish is a sure recipe for disaster. If your model is painted with realistic flat colors, the areas to be decaled must receive a generous coat of clear gloss. Some modelers give the entire model a uniform coat of gloss, but I favor applying it only to spots where decals will be applied.

In addition to a shallow container of lukewarm water, a soft No. 1 or No. 2 paintbrush, and a pad of folded paper towels, you'll need a couple of facial tissues to sop up excess water, your favorite tweezers, and a tool made by inserting a sewing needle into a piece of 1/4″ dowel, **B**.

Trim away as much clear film as possible from each marking with scissors or a brand-new modeling knife blade, but leave enough connecting film to keep letters and designs in alignment.

Work with one marking at a time. Drop the decal section into the water and give it 15 to 20 seconds for the glue that holds the film to the backing paper to loosen. Fish the decal and backing out of the water and touch them to the paper towel to blot away most of the liquid.

Apply a drop of Micro Set where the decal will go, then place the marking on the model and slide the backing out from under it, **C.** Prod with the brush to precisely position it, then blot the

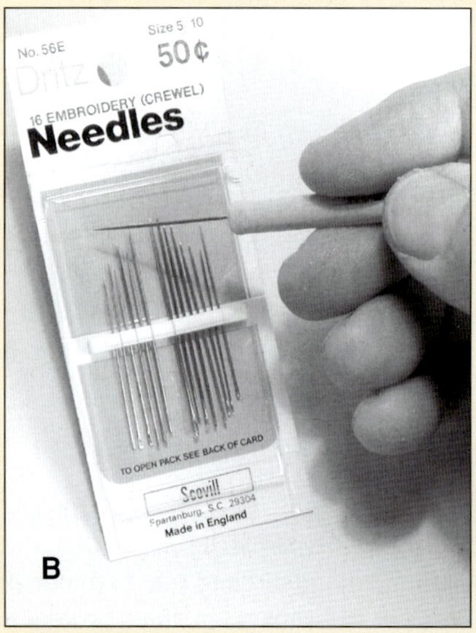

B

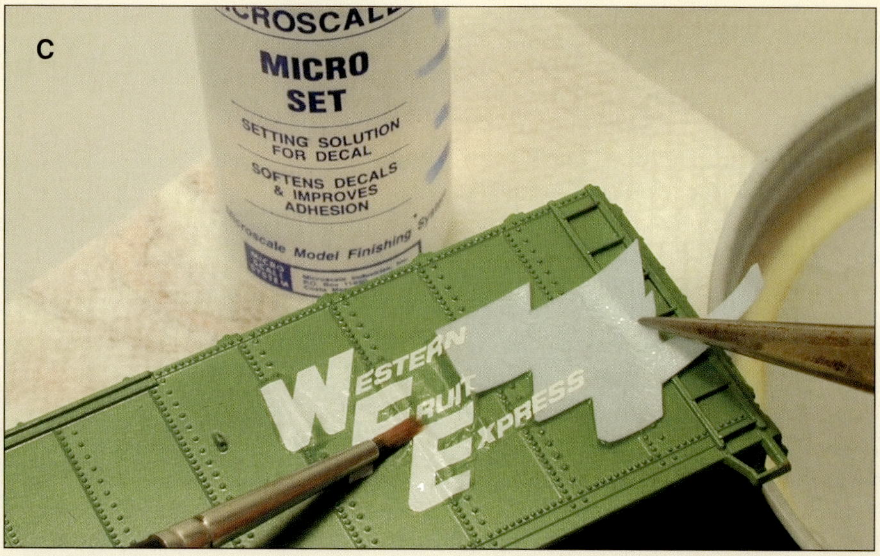

C

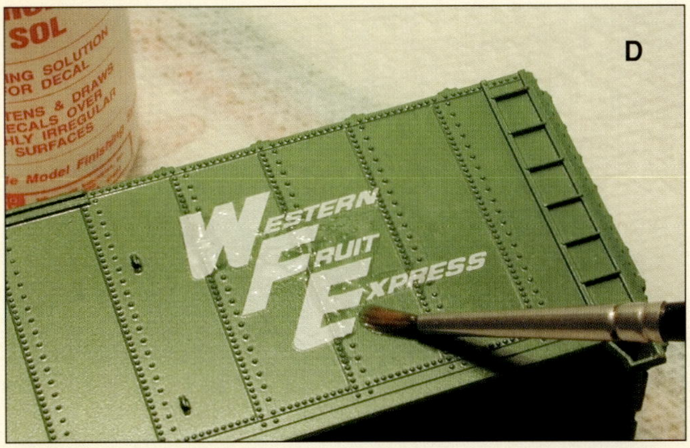

D

E

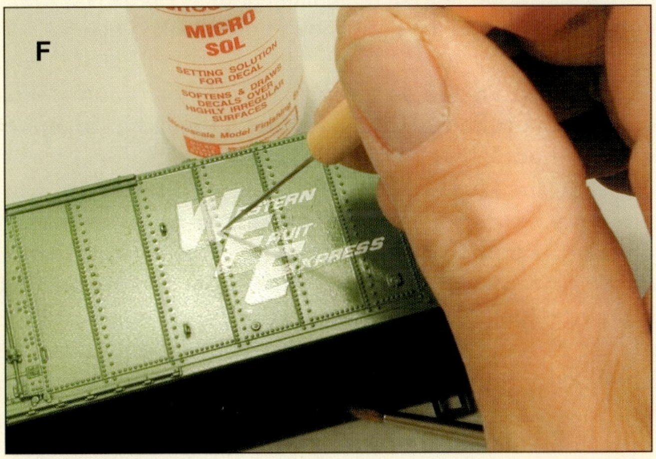

F

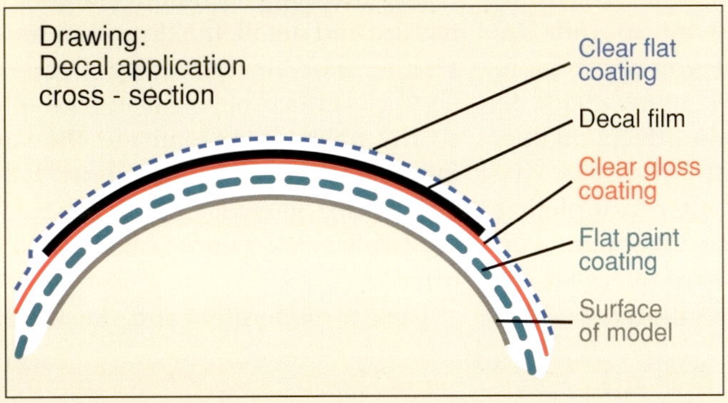

liquid away and gently press down on the decal to roll out trapped air bubbles.

While the decal is damp, apply a small mount of Micro Sol, Decal-Set, or Solvaset to the edges, **D**. This will soften the film, so it's important to avoid touching the decal from here on in. Apply enough setting agent to moisten the decal, but blot away excess so there are no drops or wet areas on top of the film.

The decal may wrinkle up, **E**, but it will flatten as it softens and conforms to the model's surface. Leave it alone. Allow the setting agent to dry, then inspect for bubbles. Prick them with your needle tool, **F,** then brush on more setting solvent and let the decal dry again.

The thickness and stiffness of decal film varies, and some brands may require up to a dozen applications of solvent to make them settle into recesses and conform to raised details like rivets. To get stubborn decals to snuggle into panel lines and scribed siding, lightly cut the film with a brand-new No. 11 hobby knife blade after your second or third application of solvent has dried, then apply more solvent.

After the decal has dried thoroughly, gently wash away adhesive residue with a moistened tissue. Allow the model to dry again, then apply a final clear flat or gloss to hide the decal film. Make the application of clear generous, so the edges of the film blend into the surface of the model.

Solving decal problems

If a decal tears or leaves a gap where it was cut or pricked, touch up the flaw with matching paint.

Old decals may fall apart when you soak them in water. Coat the decal film before application with a clear gloss coat, or with Microscale Superfilm, **G.**

Decals that refuse to stick can be applied with a diluted solution of white glue. Yellowed decals can often be saved by taping them to a south-facing window for a day or two, allowing the sun to bleach them.

Future brand acrylic floor finish, **H**, is a clear gloss that comes in handy with decals. It can be used as the gloss coat underneath decals, as a protective clear coating, and even as an adhesive for decals that won't stick. Many aircraft modelers apply problem markings into a thin layer of wet Future.

Temperature and humidity fluctuations are the enemies of decal film. Store decals in a cool, dark, dry place, in an air-tight container. — *Bob Hayden*

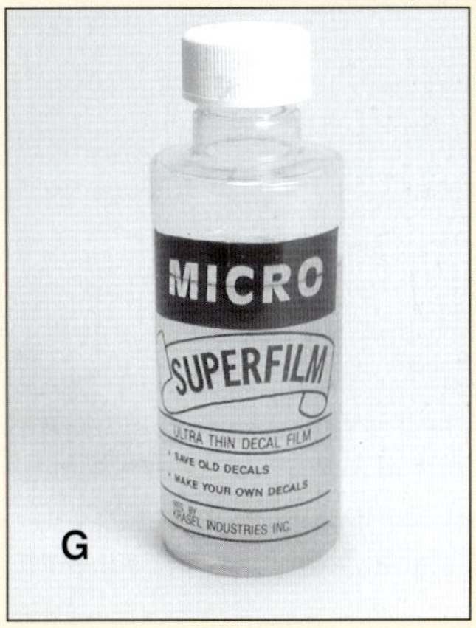

G

H

to protect the paint and markings. This flat finish can be applied by an aerosol can or airbrush, and should be allowed to dry hard — at least overnight — before you apply weathering.

The clear flat coating also adds a microscopic rough finish to the car — artists call it a "tooth" — which gives our weathering media a good surface to stick to.

This extreme close-up of an O (1/48 scale) boxcar door shows how dry-brushing works its magic. The detail on the left has been highlighted by dry-brushing with Antique White. The untreated section to the right shows how it looked before.

Pastel chalk treatments

This is the easiest weathering technique to learn, the most forgiving, and without a doubt the most subtle. The main ingredient — in fact, almost the only ingredient — is an inexpensive starter set of artist's pastels. Your first box should last roughly forever.

Grind the chalks to a fine powder, either with a big file or a swatch of coarse sandpaper. Earth-colored pastels are basically ground-up colored dirt with a binder to make them into sticks, so we can use them to represent all kinds of dirt and dust, including smoke and exhaust stains.

Brush the chalk powder onto the model with a soft brush. Apply lots, then brush most of the powder away. Work over a sheet of paper so you can reclaim the extra powder.

Keep your brush strokes going up and down, like gravity. You can use different colors, but always keep a real-world effect in mind, and always make one color predominate.

If you think you've applied too much chalk (which is almost impossible), brush it off, still going up and down. As you handle the model, try not to get fingerprints in the dust coating.

Out-of-scale fingerprints are the reason for what we do next. When you have the chalk powder pushed around where you want it, fix it in place with a clear flat spray.

The clear spray will make most of the chalk color disappear! What's left may well be exactly what you were after, and you'll want to put the model aside. But if you want more color, simply apply more chalk to gradually build up the effect you like.

Few clear fixative sprays dry absolutely flat, so after completing the chalk weathering treatment you may want to add a light overall dusting of chalk, without following with a fixative spray (especially for photos). This final coating may be susceptible to fingerprints, though.

Once you've got a couple of chalk weathering jobs under your belt you can start experimenting with different colors. The easiest to use are earth tones, but shades of brown and red oxide are excellent for simulating rust peeking through failing paint.

Painting with a dry brush!

Dry-brushing techniques highlight texture and detail, making them more visible. For dry-brushing I prefer acrylic paints. They're water-soluble, so the paint won't damage the surface below it, no matter how hard you scrub. Second, most acrylics are relatively thick, which is best for dry-brushing.

I use a moderately stiff flat water-

A simple light dry-brush application of off-white over an air-brushed slate-gray roof brings out the excellent detail in this HO (1/87 scale) injection-molded signal tower kit. Model and photo by Dave Frary.

Most dry-brushing employs a light color over a darker base. Not so here. Joe Morgan painted this 1/35 scale German reconnaissance vehicle in winter camouflage, then dry-brushed with a dark color to bring out edges and rivet detail.

color brush. For years my favorite was a Grumbacher style 626-B, No. 4. However, the last time I went to buy one of them the sticker shock sent me over to a China-bristle bright that costs about 75 cents.

The best color for dry-brushing is off white, which usually comes in a bottle labeled something like Antique White or Constitutional Parchment.

Dip only the tips of the brush hairs into the paint, then work most of the paint out of the brush by stroking it across a rag, a paper towel, or the back of your hand. When all but the last few bits of paint have been removed (it's now a dry brush, right?), stroke the brush lightly over the surface of the model.

If the brush is too dry no color will stick to the model; if it's too wet the paint will streak. But if you've removed just the right amount of color on the paper towel, paint will adhere only to the raised details, accentuating them. That's what we want.

Use vertical strokes, touching the brush to the model only on the down strokes. This will leave paint on the top edges of the raised detail. The paint makes the detail, which was there all along, appear.

A good way to think of light, overall dry-brushing is that you are adding sunlight to the model. Heavier applications of paint are good for spills or failed paint. Dry-brushing with reds and browns is especially effective for adding rust spots and streaks.

Advanced dry-brushing is often done using progressively lighter shades of the model's base color. Very subtle effects are possible with this technique.

A dark stain — this one is Payne's Grey tube oil color diluted with mineral spirits — does a great job of bringing out recessed detail and panel lines. Note the difference between the left (unstained) and right halves of this styrene crate.

Stains and washes

If dry-brushing is the way to bring out raised texture and detail by lightening and highlighting it, stains and washes are ways to do the same thing to recessed detail by darkening and visually deepening grooves and shadows.

You can use any kind of paint for stains and washes, but my favorites are flat enamels such as Testor Model Master or Humbrol, thinned with mineral spirits (paint thinner). This combination has a very low surface tension and makes the wash flow better than water-based colors, but won't attack paint like lacquer thinner. (You can substitute artist's tube oil paints for the hobby enamels.)

The proportions of paint and thinner are not critical, just use lots of thinner. Start with 20 parts to 1 part paint, then add more paint as you become comfortable with the technique. Use dark washes on most

Geoffrey Stippes and Paul Scoles worked together to scratchbuild this S (1/64 scale) railroad car ferry from styrene. Subtle use of paint and India-ink washes and pastel weathering gives the model a well-maintained yet weather-beaten look. The fancy lettering is custom-made dry transfers. Photo by Paul Scoles.

models, especially light-colored models. Flat black and brown enamels make good washes, as do Payne's Grey and Burnt Umber oil colors.

Apply the wash liberally, allowing it to run into and settle in cracks and around raised details. If the wash alters the overall color of the model too much for your liking, wipe the surface with a clean rag, or brush the excess color away with a clean brush moistened with thinner.

Once you've applied the wash to all surfaces, start wiping it off open areas with the brush. Keep stroking, bearing gravity in mind, until the surface of the model begins to dry and streak. Then put it aside to dry.

The dark wash will settle in recesses and around raised detail such as rivets, further accentuating them. (Weathered models always look more detailed than their clean counterparts!) Mix heavier washes and stains to simulate spills and heavy accumulations of dirt and grime.

If you like the initial weathering treatment on the model, try repeating it. Simply go through the basic steps of wash, followed by pastels, followed by a light dry-brushing all over again. The net effect will be to build the illusion of depth.

Underpainting

To simulate failing paint and washed-away dirt and grime, we paint the model one color, add a second color, then use thinner or abrasive to selectively strip the second color away.

Here, a stable finish is especially important; the underlying coat must be cured thoroughly so removing the second color will not affect it. The brush that you use to remove the top coat of paint should only be dampened with thinner.

To simulate weathered painted wood, apply a weathered wood color as base coat. After the base coat dries completely, paint on the finish color and use a fiberglass eraser to rub off the finish color, exposing the weathered tones.

For modeling weathered, unpainted wood, apply a wood-grain texture on the styrene before the model is built. You may then weather with thin washes similar to staining wood.

Paint on a base color slightly lighter than the final color desired using an enamel or lacquer paint. The color can range from yellow-tan to gray-brown. After the base color has dried thoroughly, highlight the surface with lighter grays and browns. Use only a very small amount of paint and dry-brush the surface lightly.

After the highlighting has dried, apply thin washes using black and brown. The washes will flow into the cracks and accentuate the wood grain. Another good wash is a solution of black and brown leather dye diluted with 50 to 100 parts denatured alcohol.

Advanced airbrushing techniques

When we use an airbrush for weathering and camouflage application, we begin to employ the tool for the tasks for which it was designed — fine-line spraying and feathering painted edges.

Fine line spraying is easy, but requires that you practice to develop a feel for the tool. Many airbrush instructions include exercises to teach the skill.

Start with a few sheets of clean white cardboard or styrene and a well-thinned color you don't plan to use for anything else. Adjust the airbrush to a low paint volume and practice spraying fine lines with the airbrush held varying distances from the surface.

Get close — as close as 1" — and experiment to see just how little paint you can apply. Then tilt the brush relative to the surface to vary the spray pattern from a circle of paint to an ellipse. Note how the tilt makes the spray tail off into a feathered pattern. Feathering edges is a strong suite of the airbrush, and this is particularly helpful for painting camouflage patterns on military subjects.

The airbrush is also especially good for applying dust and exhaust effects, because the airbrush blows the paint onto the model in a manner similar to the way the real object becomes dirty. Always thin your paint a little extra, otherwise the airbrush tip will fill up with paint and clog frequently.

Brian Ellerby simulated peeling paint on this S (1/64 scale) shed by first painting the model with a weathered-wood base coat. He dabbed thinned rubber cement on spots where the base color was to be exposed, sprayed finish color over the model, then immediately removed the rubber cement blobs with masking tape.

Subtlety is what airbrush weathering is all about. Because of that, it's better to err on the side of too little instead of allowing yourself to overdo it. Always use more than one color to avoid a homogenized look.

One of the final frontiers in realistic painting is to vary the gloss of parts of the model rather than the color. One example is applying a dead-flat dull finish on the hot and dirty parts of a steam locomotive, but a semi-gloss finish on the tender behind it. Another is to alter the gloss on individual panels of a natural-metal aircraft to call attention to the fact that they are different, contrasting metals or alloys.

Practice, practice, practice!

While none of the finishing techniques described in this chapter is difficult, all require practice. The best way to master them is to practice on an old model, unassembled model parts, or even old kit boxes. Once you become comfortable with the techniques by painting objects you'll never have to show to anyone, you can go to work on a project that's "for keeps."

RUBY'S ROOMS — A STUDY IN WEATHERING

by Brian Ellerby

Ruby's Rooms, an S scale (1/64) styrene structure, illustrates several basic weathering techniques. The model is based on a rooming house in Tonopah, Nevada.

All-styrene construction

The siding is Evergreen 5-1/4″ Novelty Siding with a wood grain surface. The vertical lines on the walls mark the stud locations. Board ends are randomly scored at the studs to simulate individual boards.

The rear wall and upper part of the left wall are more heavily wood grained and have nail holes marked with a sharp scriber. The siding below the porch roofs does not have the nail holes or heavy weathering because these walls are protected from the weather. All surface texture was added before the window and door openings were cut and the walls assembled.

Inside corner braces are .060″ square strips. A ceiling of plain .040″ sheet is glued permanently in place to brace the walls and keep the corners square. Removable floors provide interior access for painting and detailing. The floors are held in place by strips glued to the inside of the wall at floor level.

Individual styrene shingles were applied to the sub-roof prior to attaching the roof panels, and the roof and roof trim were installed before painting to allow a strong styrene-to-styrene bond between the floor and walls.

Painting and weathering

Most weathering was completed prior to final assembly of the porches and porch roofs. Windows and doors were prepainted a contrasting color before installation. I prefer to install corner trim before painting, even though the trim must be masked to paint the model. Overall, this is faster than adding painted trim strips to the completed model.

The entire model was first airbrushed with a basic gray-brown mix of Floquil Railroad Colors to simulate weathered wood. After the base coat had completely dried, Floquil Antique White was lightly sprayed on the walls. Additional paint was sprayed on areas protected from the weather by the roof overhang.

Rubbing a fiberglass eraser over the walls exposed the gray-brown base color. A lighter gray (SP Lettering Gray) was sprayed on the porch roofs and decks and rubbed off using the same fiberglass eraser. This technique is fast and easy, but be careful not to overdo the effect — or you'll be back to an "unpainted" weathered-wood colored model!

Porches, roofs, and final treatments

Individual styrene strips were used for the porch and porch roof. The strips were heavily wood grained before cutting to the finished length. Each porch roof and deck was completed as a separate subassembly.

Floquil Depot Olive was used on the trim, windows, and door moldings. The tarpaper on the porch roof is masking tape applied over the prepainted planking. After painting with Grimy Black and overspraying with gray-brown, the tape was peeled up and trimmed with a sharp razor blade.

The corrugated metal roofing was painted with a mix of Roof Brown, Rust, and Black to simulate rust. Humbrol Sea Grey Medium was then dry-brushed on the roofing to simulate weathered galvanized metal.

The final step was a thin wash of black leather dye in alcohol brushed over the entire model, particularly the shingles. Details yet to be added include interiors, moss on the roof, and, of course, Ruby sitting in her rocker on the porch.

This out-of-the-box HO (1/87 scale) locomotive was weathered with light applications of rust, earth, and grime colors. The most interesting effect is the vertical streaking achieved by airbrushing through a simple cut-paper mask.

Chapter 10

Scratchbuilding Projects and Techniques

THERE IS NO SUCH THING as a truly typical styrene scratchbuilding project. Building models from scratch is something you learn slowly by doing, rather than by sitting down and taking lessons.

For that reason, our final chapter consists of a couple of sidebars on advanced modeling techniques and several scratchbuilding case studies. By watching as experienced modelers tackle the challenges inherent in wildly different subjects, you can gain insight to be applied to your own projects.

With one significant exception, these projects employ styrene to represent metal surfaces, which is one of styrene's strong suites. Just as important, though, is a principle mentioned in the very first chapter of this book: Don't hesitate to combine styrene with other materials when it will make for a better model.

One example of this principle in action is delicate (and therefore, easily damaged) assemblies. Hand grabs and railings figure prominently in the projects in this chapter, and you'll find that the modeler often substitutes brass or other metal wire for styrene where the parts will be vulnerable.

Happy modeling – with styrene!

David Merriman scratchbuilt this 29″-long Star Trek *Enterprise* almost entirely from vacuum-formed styrene sheet. The lighted hangar bay was completed and painted before being integrated with the rest of the model.

This detailed HO (1/87 scale) model of the huge Barrett Bridge on the Yosemite Valley Railroad is the work of Jack Burgess. The 44″-long model was scratchbuilt almost entirely of Evergreen styrene sheet.

Al Kalbfleisch built most of this Union Pacific coaling tower from basswood, but made the supporting structure and staircases from styrene. He turned to brass wire, however, to fabricate the delicate handrails on the HO (1/87 scale) model. The inset photo shows how he used clear styrene to make gusset plates with embossed rivet detail.

A

B

STYRENE SCRIBING TECHNIQUES

by David Merriman

Often a model requires creation of special tools. Here, an SSBN 640-class missile submarine (its hull and sail are fiberglass; the superstructure is vacuum-formed styrene sheet) is being scribed using a special template.

Scribed grooves in the surface of the styrene superstructure represent the 16 identical missile hatches on this model. Photo **A** shows me using a simple plastic scribing template as I engrave a hatch line. The template has additional openings for scribing line lockers, reversible cleats, and main ballast tank vents.

Curved scribing templates

Forming scribing templates for compound curves like the bow section of this submarine can be done by using the model as a plug to shape heated plastic. It works here because the fiberglass hull can withstand the heat — I would not attempt this if the hull were styrene!

In **B**, a square of .020˝ styrene has been tacked to a plywood backing piece with a round hole. After heating over a hot plate, the plastic was pushed onto the sub's nose, and the circular cutout assured a tight, faithful impression of the bow contour.

The locations of the square torpedo tube shutter doors were laid out on the now-cone-shaped template, and the square holes opened with drills and files, checking the dimensions frequently.

Mounted on the bow, **C**, the scribing template is ready for use. The scribing tool is dragged with light pressure over the surface of the model, controlled by the edges of the cutout. Succeeding passes are made with increasing pressure until the engraved line is of the desired depth, **D**.

Round jeweler's files make excellent blanks for forming scribing tools. I shape the tool to a point, then, holding the point at a right angle against a sharpening stone, slightly grind off the tip, producing a microscopic flat.

Make the scriber from high carbon steel if the job entails working fiberglass, acrylic, or some other tough plastic.

Fixing mistakes

To fix slips or tool marks I remove the template and smear automotive filler putty over the entire scribed portion of the model. Quickly, the scriber is run freehand with light pressure through all the engraved lines to chase out the still-wet putty.

What putty remains on the model will have filled the imperfections. When dry, the puttied area is wet sanded with 400-grit paper, and the scriber is used again to chase out sanding dust in the engraved lines. The area is given a light spray coat of primer and inspected. Putty is reapplied to remaining flaws and the process repeated.

Here's another custom heat-formed scribing template taped atop the sail of a submarine model, **E**. From left to right: 400-grit wet-or-dry sandpaper, file, jewelers file, X-acto knife, pencil, and scriber (a sharpened file point mounted in a pin vise). Most of these tools were used to open and shape the scribing template cutouts.

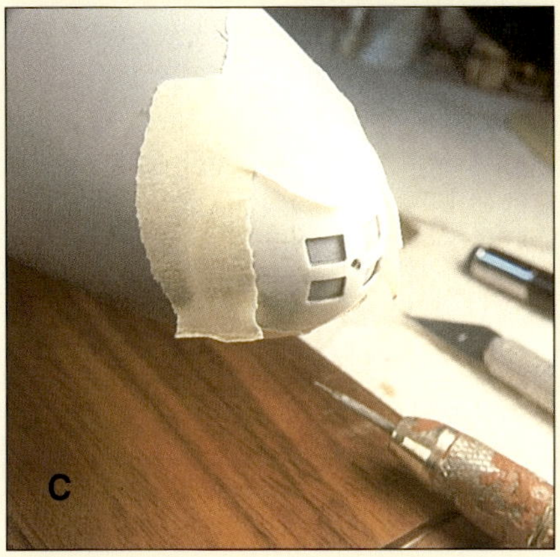

C

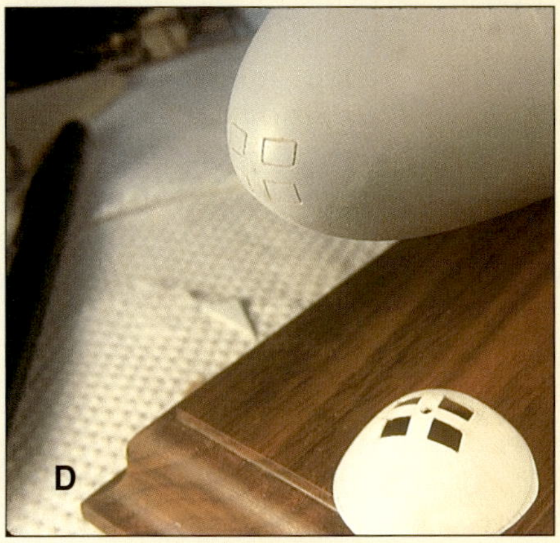

D

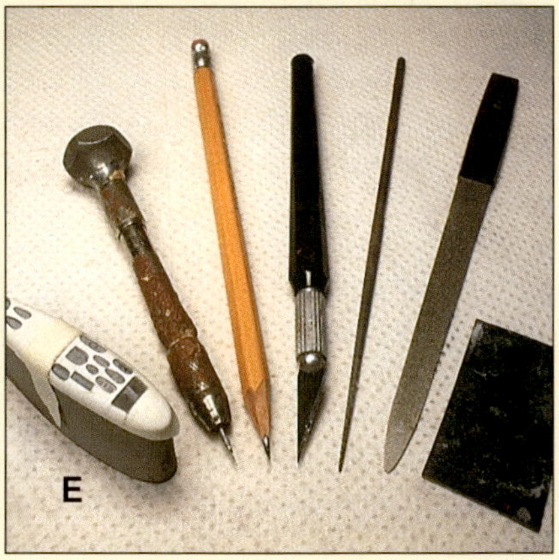

E

VACUUM-FORMING SIMPLE PARTS AND CANOPIES

by Pat Hawkey

One of styrene's greatest assets is its pliability. This is especially true when heated. Heat-softened styrene can be pulled over shapes or sucked into cavities to reproduce almost any shape, including complex curves.

Back in the early '60s Mattel introduced the Vac-U-Form machine. Sold as a toy, it provided youngsters with a hand-operated vacuum pump, heating element, styrene sheets, and molds to make their own plastic toys. Like most hobbyists, I found mine at a church rummage sale long after the animal was extinct and paid about 50 cents for it. Thousands of them were made, and they still surface on the yard-sale circuit.

A

B

C

Landing gear doors

I made master forms for a pair of "fast-pack" landing gear doors for a 1/48 scale DC-3 from 2-part epoxy putty, **A**. This material cures to a granite-like consistency, impervious to the heat of the plastic that will be drawn over it.

The master on the right is upside down, showing a .030″ spacer that will keep the master slightly elevated. This ensures a sharp bottom edge where the heated plastic will be sucked into the gap between the master and the vacuum platform.

To make the duplicates, I loaded my trusty old Mattel Vac-U-Form machine, **B**, with a 3″ x 3-5/8″ sheet of .020″ plastic. The heating element has done its job and the plastic is shiny, sagging, and ready to go.

The thickness of the plastic you start with will not be the thickness of your finished part. As the heated sheet is stretched over the master it thins over the high spots. A good rule of thumb is to decide ahead of time how thick and rigid you want your finished part to be and select styrene .010″ thicker.

The clamp assembly is quickly flipped over to the vacuum platform, the suction handle is given a couple good pumps, and if you don't have any air leaks, here's what you get, **C**. (A close look will reveal the door on the right is flat while the door on the left is on edge. Being more focused on taking pictures than obtaining good parts, I forgot to secure the masters to the vacuum platform with a dab of white glue to keep them from shifting.)

The reverse side of the cooled sheet has the two masters firmly entrapped in the styrene, **D**. I roughly cut the doors free with scissors, then trimmed off the remaining excess with a new No.11 blade, **E**. It's easy to slip the point of your blade between the new part and the master and trim by just riding that line.

A few swipes with an emery board or a sanding stick to finish the cleanup, and I have a good pair of doors, **F**. Photo **G** shows where they ended up.

Vac-forming replacement canopies

Like many model plane builders, I wanted a Mattel machine to make replacement canopies. I believed that getting my hands on a Vac-U-Form would simply and quickly turn modeling night into day. Not quite. Here's what I've learned about making canopies through trial and error — and lots of wasted clear plastic.

1. Clear styrene does not work.

Instead of softening and sagging when it's heated, clear styrene shrinks and pulls itself out of the holding clamp. Acetate and Butyrate sheet work, but I've had bubbles form in both as they heat up. Don't know why — something about moisture absorption, I suspect. There's also Polycarbonate and Vivak you can experiment with. All these materials are soft

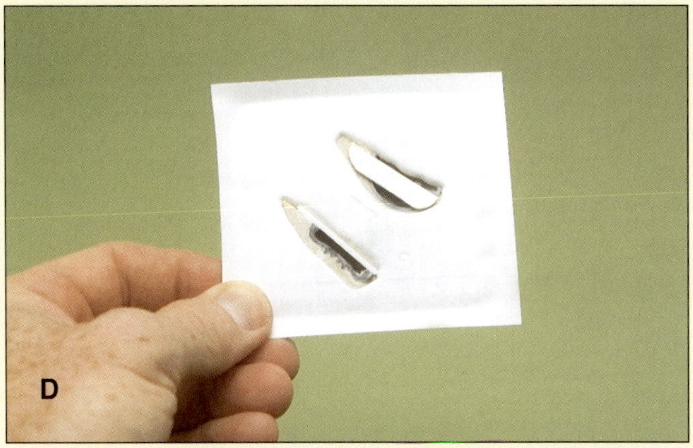

D

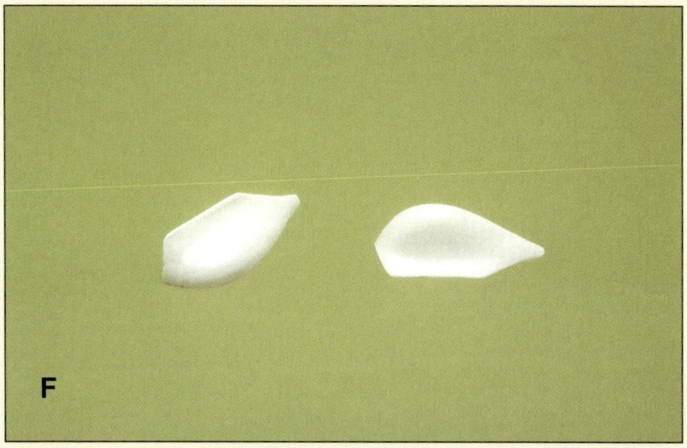

F

E

G

and easily scratched. A scratch on your unheated sheet is not going to go away after you've made a canopy out of it.

2. Thin is what you're after, but thin also means flimsy.
The new canopy should be thick enough to provide rigidity in handling. You may have to compromise between scale thickness and practicality.

3. You may ruin the kit canopy.
Heating anything but the thinnest sheet stock, then sucking it down over a kit canopy used as a master, can melt and ruin the kit piece. The safest thing to do is make a master out of epoxy putty, a resin casting, wood, or other material that will stand the heat.

4. Your canopy may be oversize.
The canopy you form will be bigger than the master by the thickness of the material. This is no problem if the canopy will be slid back or raised, but if you're looking to have it closed, a vacuformed replacement based on an unmodified kit piece will be bigger than the hole it's expected to cover. And take it from me, in this case, bigger can't be made to work.

5. Fastening the canopy to the model can be tricky.
None of the clear plastics that I've used react to cement for styrene. While white glue will hold clear parts in place without crazing, unless you've got a flawlessly fitting piece, the end result with white glue tends to look like the canopy was a last minute add-on.

H

On real airplanes the clear parts usually blend smoothly with the not-clear parts. The way to achieve this on a model is to blend the canopy bottom framing into the rest of the model with body putty.

The only way to hold the clear piece in place during this blending operation is with super glue, but super glue tends to frost the surfaces of clear material. (Carefully applying tiny amounts and allowing them to dry before adding more tiny amounts can keep this side effect to a minimum.) If you can get to the frosted area, you can polish it clean. If you can't get to it, well.

For me, pulling out my Vac-U-Form machine to make a canopy has become a method of last resort. Despite all of this, photo **H** shows a successfully vacuum-formed clear canopy on an Otaki/Arii 1/48 scale P-47.

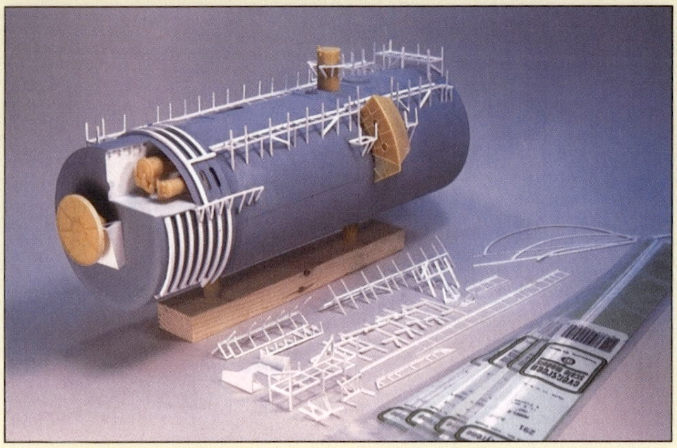

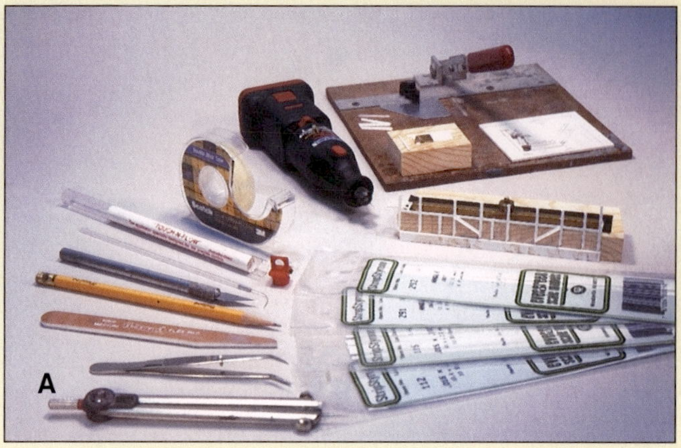

A

CASE STUDY IN STYRENE: SUBMARINE SHOCK-TEST VEHICLE MODELS

by David Merriman

I can think of no better example of using Evergreen structural shapes than the model pictured here. This 1/96 scale model represents a unique subject: a section of submarine hull built for underwater testing of submarine systems in a high-shock environment. This vessel is exposed to underwater explosions of considerable power detonated at short ranges.

I was commissioned to build three of these models, each requiring significant amounts of strip, angle, H, and T-section structural shapes. Photo **A** shows the tools and consumables I used. Three items are worthy of further comment: Scotch Brand Double-Stick Tape; The Chopper from NorthWest Short Line; and the Touch-N-Flow adhesive needle applicator produced by Creations Unlimited Hobby Products.

The tape permits me to temporally mount subassemblies to check alignment and correct part fit. The Chopper makes exact duplicate cuts of any desired angle to produce identical parts. The Touch-N-Flow permits better flow control of solvent adhesives than the traditional paintbrush.

The hull of each A/B-1 is a length of acrylic tube, forward and after bulkheads are sheet styrene, and many of the appendages (including the inserted torpedo bay) are cast urethane resin, **B**. The deck and hull fittings are spin-cast white metal. Most of the walkway and platform structures were assembled in fixtures from Evergreen structural shapes and sheet.

I used a pen-loaded surface gauge to mark off the radial and longitudinal lines that locate the items attached to the cylindrical hull. To mark the radial lines indicating where the external frames would be I rotated the hull within its working fixture.

External frames: A special challenge

The T shapes needed to represent the external frames at one end of the model presented a special problem. I could not use a commercial item, as it would resist bending and buckle. The solution was to build the frames in place on the hull from two pieces of strip stock.

I started by working the hull-mounted strip into the desired curve, then gluing it to the hull. A simple bending tool made from a wood dowel was used to precurve the strip to the desired shape.

Applying pressure with my thumb to bend the strip as I pulled the strip down introduced the stress needed to start the piece on its way to adopting the needed radius, **C**. Repeated passes through the tool produced a tighter radius.

Photo **D** shows the strip stock precurved to the approximate radius of the hull. The photo also shows one end of the hull holding fixture, which permitted me to rotate the model.

I used thin super glue to bond the strip to the hull. To achieve fine control of adhesive placement, I transferred several drops to the bottom of an inverted glass bottle. From there I used a pointed stick to transfer precise amounts to the work, **E.**

Holding the curved strip to the hull with my thumb, I applied a small drop of super glue to fix one end in place. The hull was then rotated and the other end of the strip pulled tight before a second drop of glue was applied to secure it. Super glue was then liberally flowed on to bond the strip to the hull along its entire length.

I then placed the second piece of strip

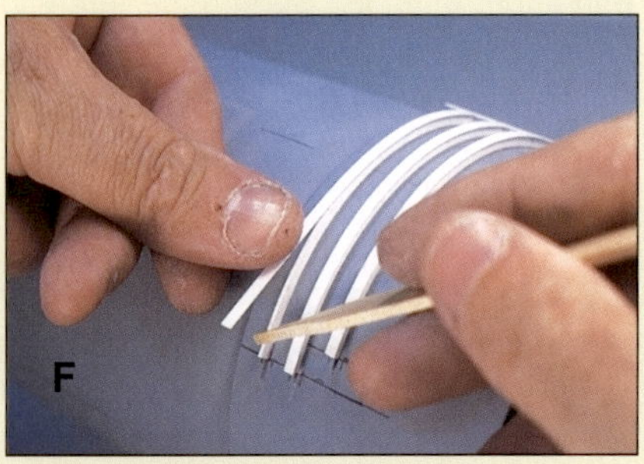

F

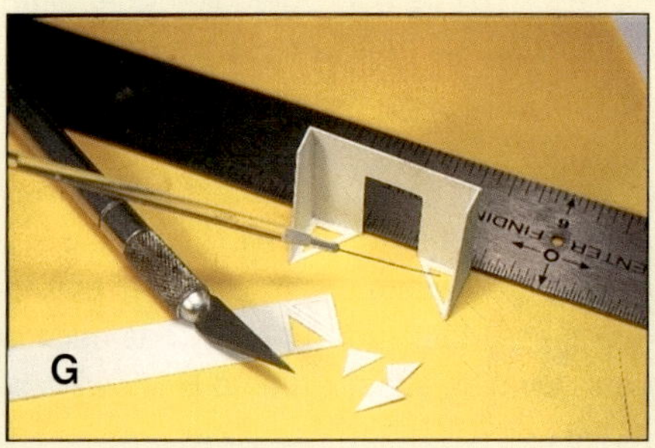

G

H

B

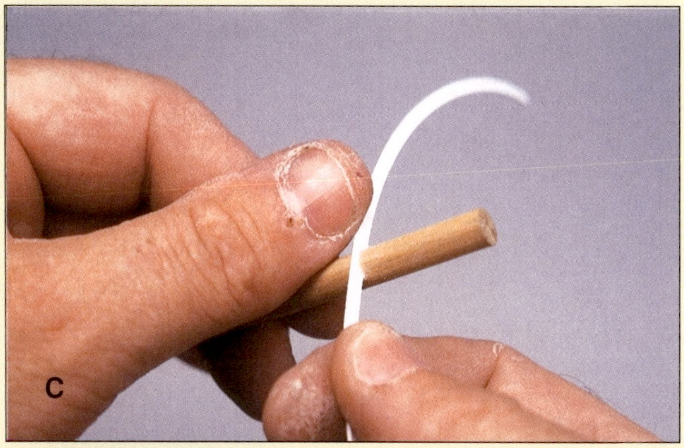

C

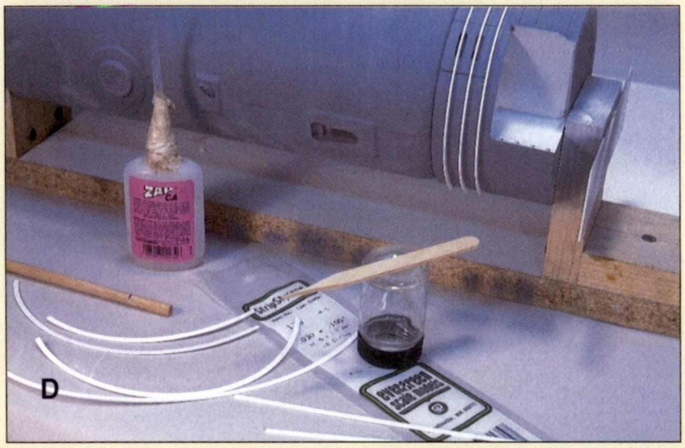

D

E

stock over the first at a right angle to form the characteristic T-section external frame, **F**.

Wicking away excess solvent

One trick to avoid solvent cement from getting under the work and sticking it to your work surface is to lay down, with the aid of double stick tape, a piece of absorbent cardboard on the bench top, **G**. Excess adhesive is quickly drawn into the paper. Of course, as soon as an item is glued it must be scooted away a bit so solvent absorbed by the paper won't slowly mar the finish of the parts touching the table.

Here I'm using a gravity/capillary feed needle applicator to precisely deliver solvent to a right-angle truss between two pieces of sheet. Note that I made the triangular piece from a single piece of sheet — this is a whole lot easier than trying to jig three little strips of plastic together!

I temporarily fitted styrene platforms to a cast cylindrical access trunk with the aid of double-stick tape, **H**. These channel-section pieces were solvent glued to the underside of a platforms. After the bonds had hardened, the platforms were removed while the resin access trunk was sanded, primed, and painted.

Special assembly, sanding, and drilling fixtures were made to help cut and fabricate the walkways, boat fenders, stanchions, and platforms, **I**.

On the walkway assembly jig, **J**, a brass strongback holds the pre-assembled upper walkway against the top of the jig. (Sandpaper glued to the jig prevents the work from shifting as stanchions are glued in place.) The channel stock stanchions rest against the vertical face of the fixture.

The white plastic bar at the base of the vertical face sets a uniform height at which the stanchions intersect the horizontal channel forming the upper portion of the walkway.

Painted white overall, this model was weathered with chalks and water-soluble paints, **K**. Black dry-transfer lettering represents the draft numbers and other markings.

I

J

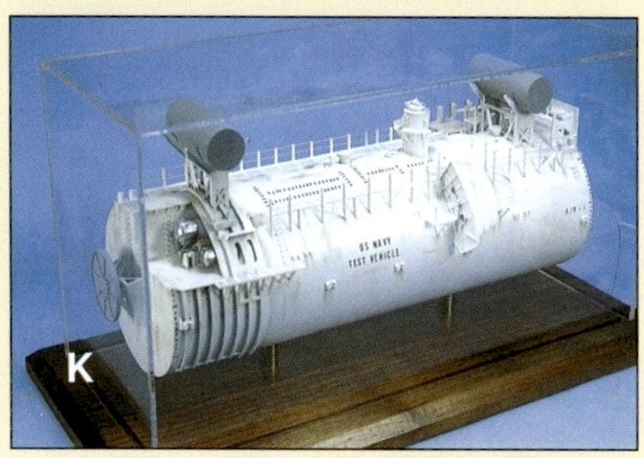

K

CASE STUDY IN STYRENE: THE LITTLETON POWER COMPANY PLANT

by Al Kalbfleisch

The idea for the this HO (1/87 scale) model came about after seeing an old M.I.T. text book my dad used back in 1929. The model was based on the same function and flow diagram as a Dodge Brothers plant in Detroit. It has an internal coal storage bin, twin boilers, closed system turbogenerator, stokers, appropriate internal plumbing, and lighted control panels.

All of the water and steam plumbing is styrene rod and the overhead crane has a control cab, trolley frames, chain fall motor and rail surface details made of sheet and strip styrene, **A.** The building shell has styrene fascia and corrugated roof, **B.**

This model was hard on my spare parts stash, as I used kit components as the basis for the interior machinery, **C**. There was plenty of opportunity for adding details using strip and structural shapes.

Lighting and sound effects

The lights are on four circuits: a flashing red beacon on the stack, the lights used to backlight the interior, hanging lights internal to the plant, and control panels. This arrangement allowed balancing the brightness for the desired visual impact.

The control panels consist of internally lighted sheet styrene boxes with colored transparencies representing the switches and dials on the front, **D**. With the exception of these lights and those attached to the furnace floor, all lighting was carried by the outer building shell.

All bulbs were installed with resistors in series to reduce heat output and extend bulb life. Heat is important when installing bulbs in styrene enclosures — make sure they do not run hot to the touch!

The building has a sound system that includes boiler noises, pop valves, phone buzzers, loud conversation, even slamming doors. External coal storage

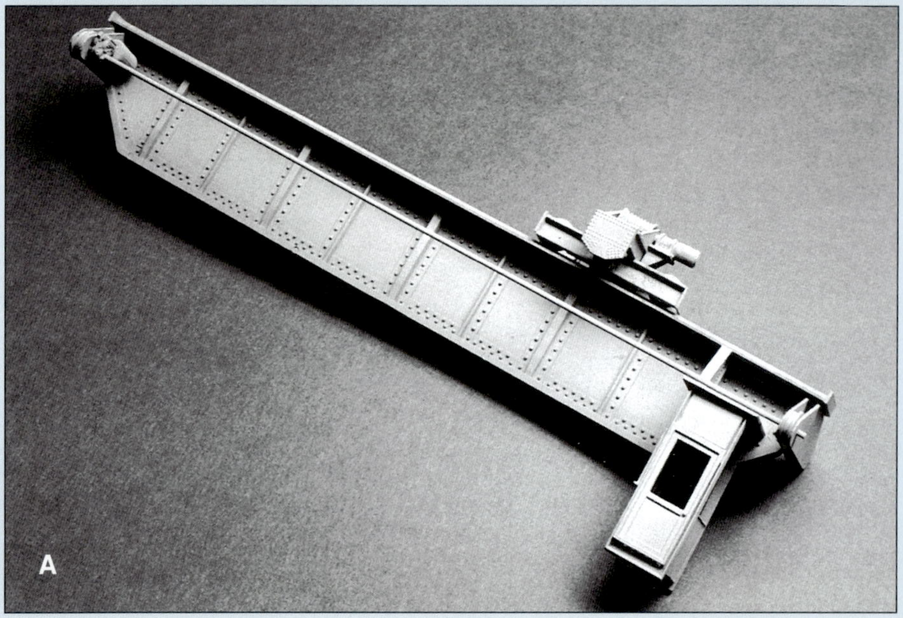

A

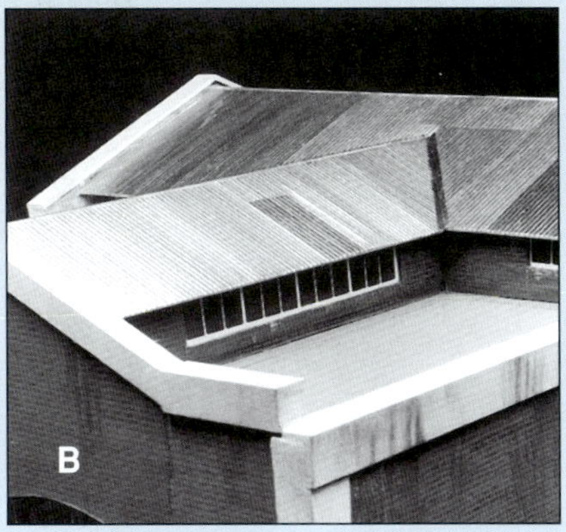

B

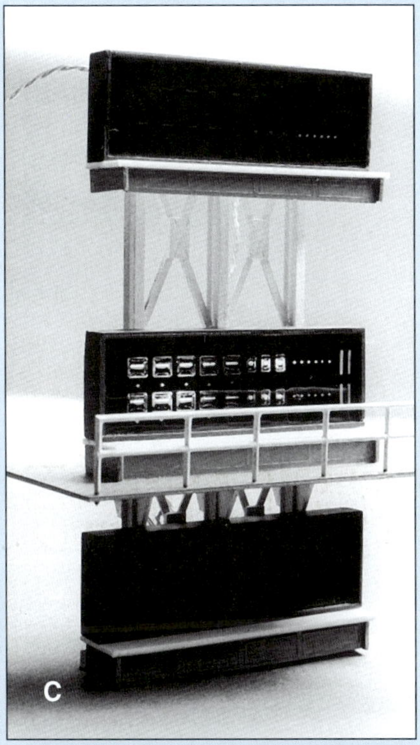

C

D

allowed for a small switching yard and provided an opportunity to build a large bucket crane to service it.

CASE STUDY IN STYRENE: SCRATCHBUILDING EL PORTAL STATION

by Jack Burgess

This HO (1/87 scale) model demonstrates the versatility of styrene. The prototype El Portal station at the gateway to Yosemite National Park was designed in the rustic style popular in the Western national parks at the time.

The building used 8″-10″ rough-hewn poles for the main supports, and the walls between them were sheathed in patterns of Incense Cedar bark. Trim at the top of the station was accented with straight branches cut to fit.

The model is constructed of .040″ sheet styrene with .060″ styrene for the floor. The horizontal and vertical exterior poles were duplicated with .062″ and .093″ styrene rod. Since half-round rod was not available at the time, I milled Evergreen styrene rods into a half-circle shape using a Dremel cutter in a drill press.

Adding rustic wood grain

While I usually don't add grain to HO scale models (you can't see grain on most well-maintained surfaces at this size), I did for this model because of the rough-hewn materials. Coarse emery cloth was dragged across the surface of each pole to roughen them before they were bonded to the shell.

I used HO 2x4 and 2x6 styrene strip to duplicate the cedar bark. The edges were slightly rounded with an X-acto knife blade (both to duplicate the prototype and to make the joints between individual pieces stand out). Grain was then scratched into the surface with emery paper. Random knotholes were added by twirling an X-acto knife vertically on the pieces.

Rather than try to cut and fit each piece of bark into the spaces between the poles, I bonded the treated pieces to .005" styrene sheets. Small squares of these sheets were then cut and bonded in place.

Modeling the trim around the top of the building employed a similar technique. I cut .025 styrene rod over-length and bonded it to strips of .005″ sheet. Sections of this trim were cut to fit the openings between the horizontal and vertical poles.

Styrene + brass = strength

The roof of the real station consisted of log trusses with solid sheathing over the eaves and skip sheathing over the interior of the building, plus cedar shingles. A special feature I planned into the model was a cutaway roof to show off the fully detailed interior. **A**.

I knew that the large eave overhang would be subject to unacceptable warping due to the thin material needed to replicate it correctly. Therefore, once the sheathing (.020″ scribed styrene with the scribing down) was bonded to the completed styrene trusses, .010″ brass sheets were cut to fit the roof, soldered together to add strength, and bonded to the top of the styrene roof with non-flammable contact cement. This gave the roof the needed rigidity.

After painting the model, bond paper was laminated to the brass with spray adhesive and Campbell brand paper shingles added.

A

B

A

CASE STUDY IN STYRENE: CN'S SPECTACULAR CISCO BRIDGE

by Doug and Jackie Hole
Photos by Don Weixl

Just west of the small town of Lytton, British Columbia, Canadian Pacific and Canadian National change sides of the Fraser River at a place called Cisco. CN's massive orange and black steel arch bridge crosses the muddy Fraser high above CP's more delicate overhead truss, which seems to disappear into a tunnel on the opposite side of the river.

The HO (1/87 scale) diorama shown here and on page 68 is 12´ long by 2´ wide, and breaks down into two 6´ sections. The bridge is 809 scale feet long, and includes 52 modified plate girder spans. Evergreen I beams, angle cross bracing, and gusset plates have been installed according to the bridge plans.

Approach towers

The support towers, **A**, are constructed from Evergreen strips. A few odd-sized strips were cut from sheet styrene using a steel straightedge, an X-acto knife, and a single-edge razor blade.

A jig was used to hold the parts in position while working on them, and for alignment while gluing. Rivets were embossed where needed on gussets and beams using a NorthWest Short

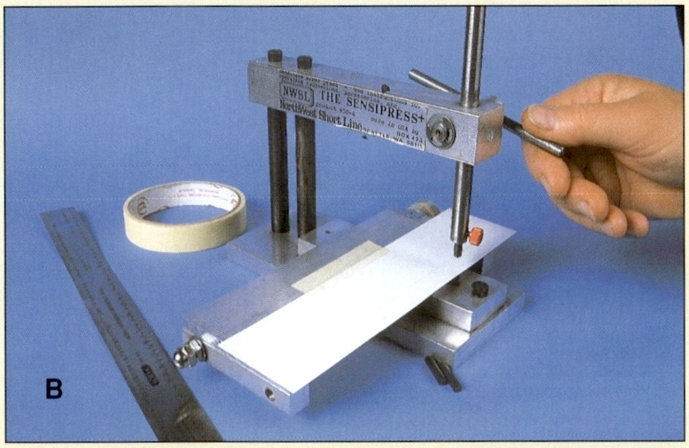

Line Sensipress and Riveter, **B**.

The towers are fastened to the concrete pads with 00-90 brass nuts, bolts and washers.

Arches and cross bracing

All of the major arch beams are scratchbuilt from Evergreen sheet and angle, **C**. Some of the internal cross bracing and latticework was created by stripping and modifying Central Valley girder material. This turned out to be more work than it was worth, so the majority was scratchbuilt out of Evergreen styrene.

The arch bridge was built in place, just like its prototype: The arches were started on opposite sides of the river and progressed toward the center, section by section, until they met. Two large custom-made brass bolts hold the two arch sections together. Special wrenches had to be made to tighten the inside nuts.

Walkways

The top walkways and girder and arch inspection walkways, **D**, are kitbashed from Tichy open grate platform kit parts. The sections are attached to styrene angle strips that run the full length of the bridge.

Vertical walkway support angles are attached to the outside angle braces of the girders. A walkway safety hatch cover hinges back for ladder access. Four post-supporting angle braces with turnbuckles are located at each end of the walkways.

Styrene channel parallels the walkways and runs the full length of the bridge. The channel angles into the ground at each end of the structure and appears in the equipment shed on scratchbuilt equipment racks.

Styrene birds, too!

The people, animals, fish, trees, and birds, **E,** were all created by Jackie Hole. After designing the basic body patterns and transferring them to styrene sheet and rod, Jackie builds up wing and tail thickness with smaller overlays.

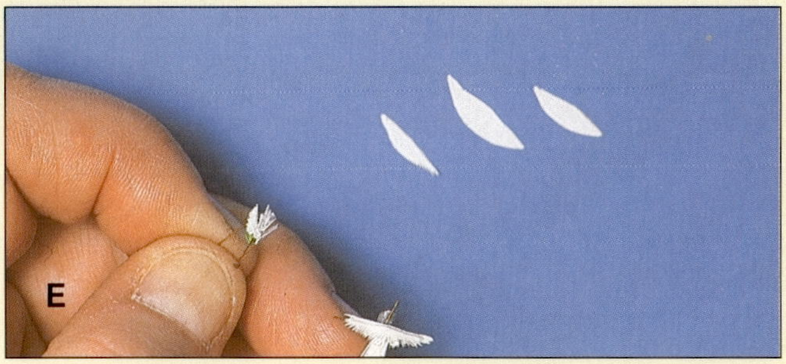

The styrene body parts are assembled with liquid cement and super glue. Fine brass wire is used for necks and legs, and for attaching the finished birds to the diorama. Heads, legs, and feet, and fillets where the wings join the body are 5-minute epoxy.

CASE STUDY IN STYRENE: SCRATCHBUILDING USS *COWPENS* IN 1/96 SCALE

by Al Kalbfleisch

My favorite project of all time was scratchbuilding this large ship model. In order to detail it to the level I wanted, I had to develop many new capabilities, including photoetching, vacuum forming, using a router, soldering metal while it's attached to plastic, making custom decals, low speed drilling, fabricating handrails and ladders — just to mention a few.

There was never any doubt about the requirement to use styrene to reproduce the ship's metal construction. Photo **A** shows the basic core of the model. The hull shape was taken from a 1/350th plastic model of USS *Bunker Hill*, and plans from The Scale Shipyard for USS *Ticonderoga*. I also had an extensive prototype photo file.

An adhesive called "Goop"

Each basic deckhouse shape was constructed from wood, smaller than required by the thickness of the styrene to be applied. That way, the final surface dimensions matched the real ship.

The outer styrene skin (up to .080″ thick) was attached with a hardware-store adhesive called Goop. No thickness less than .060″ should be applied with this cement, because of the possibility of the solvent from the Goop penetrating the styrene and softening or warping it.

A

Painting as you go

By the time the first coats of paint had been applied, the model had a lot of sheet and strip styrene on board, **B**. I feel the main reason to paint as you go is to continue inspiration, but as construction continues details become increasingly complicated and difficult to reach to paint later.

I vacuum-formed parts for several subassemblies from .005″-.007″ styrene, including the vent housing between the funnels and the control house on the flight deck. The helicopter hangar doors and storage containers aft of the bridge are styrene-covered blocks. All deck overhangs were detailed with strip styrene structural ribbing.

The completed model is on display at the Cowpens National Battlefield in South Carolina, along with another appropriately scaled model of the previous USS *Cowpens*, which served in World War Two.

B

A LOW-SPEED, LOW-POWER DRILL FOR STYRENE

by Al Kalbfleisch

Styrene is a wonderful material — it can be cut, sawn, scraped, filed, stretched, welded (chemically), and drilled. It's this last process which I find the most challenging. Faced with the need to make several hundred mounting holes for handrail stanchions, a pin vise would not do the job.

Standard motor tools and drill presses aren't the solution either, because they're too fast and too powerful. The drill heats the plastic, and what you pull out of the hole is a plug several times larger than the hole you need. Often, the part is ruined.

Two model railroad friends came up with the perfect solution. Don Kassler designed a low-power drill with a 12-volt motor, and Harold Horner built one for me! I confess I helped a little: I added a foot switch and train power pack for speed control, **A**.

This homemade drill has allowed me a great deal of freedom and saved a lot of time. It's the best way to drill plastic, and should be a standard tool in any styrene modeler's workshop. I've provided a sketch, **B**. Due to its intentionally low power, the tool can handle only small drills, so the trick is to enlarge holes a bit at a time.

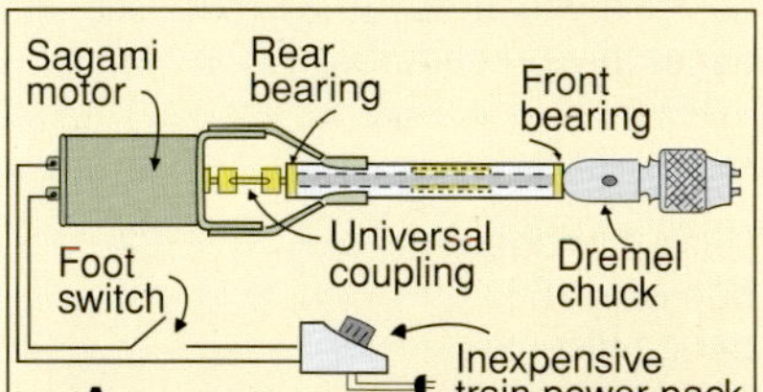

A

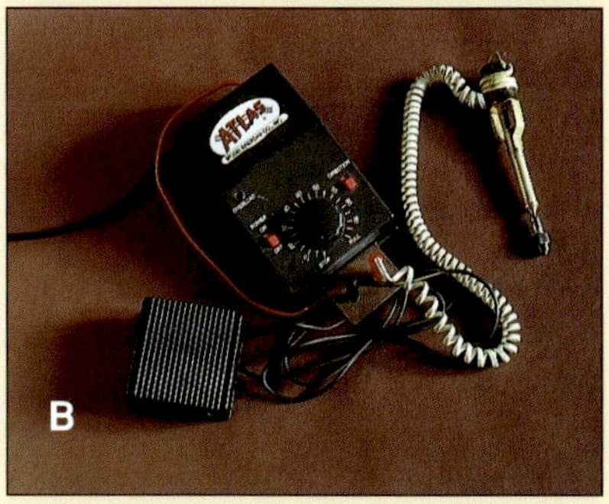

B

SOURCES

Most of the products mentioned in this book are available at well-stocked hobby shops; many others can be found at art, craft, and hardware stores. The contact information below is for companies with especially useful or occasionally hard-to-find items.

Creations Unlimited Hobby Products
(Flex-I-File, Touch-N-Flow applicator)
4318 Plainfield Ave. NE
Grand Rapids, MI 49525
(616)-363-1242

Grandt Line Products
1040-B Shary Court
Concord, CA 94518
(925)-671-0143

Micro-Mark (modeling tools)
340-2795 Snyder Ave.
Berkeley Heights, NJ 07922
(800)-225-1066

MV Products (headlight lenses)
Available from Wm. K. Walthers
P.O. Box 3039
Milwaukee, WI 53201-3039
(414)-527-0770

North Coast Hobbies (spray booth)
P.O. Box 31
McKees Rocks, PA 15136-0031
(412)-331-9970

NorthWest Short Line
Box 423
Seattle, WA 98111-0423
(206)-932-1087

Scale Shipyard,
Warship Hulls Unlimited
5866 Orange Ave., Unit 3
Long Beach, CA 90805-4146
(562)-428-5027

Waldron Model Products
(punch and die sets)
P.O. Box 431
Merlin, OR 97532
(541)-474-3226

INDEX